Deep Belief Nets in C++ and CUDA C: Volume 3

Convolutional Nets

Timothy Masters

Apress®

Deep Belief Nets in C++ and CUDA C: Volume 3: Convolutional Nets

Timothy Masters
Ithaca, New York, USA

ISBN-13 (pbk): 978-1-4842-3720-5   ISBN-13 (electronic): 978-1-4842-3721-2
https://doi.org/10.1007/978-1-4842-3721-2

Library of Congress Control Number: 2018940161

Managing Director, Apress Media LLC: Welmoed Spahr
Acquisitions Editor: Steve Anglin
Development Editor: Matthew Moodie
Coordinating Editor: Mark Powers

Cover designed by eStudioCalamar

Cover image designed by Freepik (www.freepik.com)

Distributed to the book trade worldwide by Springer Science+Business Media New York, 233 Spring Street, 6th Floor, New York, NY 10013. Phone 1-800-SPRINGER, fax (201) 348-4505, e-mail orders-ny@springer-sbm.com, or visit www.springeronline.com. Apress Media, LLC is a California LLC and the sole member (owner) is Springer Science + Business Media Finance Inc (SSBM Finance Inc). SSBM Finance Inc is a **Delaware** corporation.

For information on translations, please e-mail editorial@apress.com; for reprint, paperback, or audio rights, please email bookpermissions@springernature.com.

Apress titles may be purchased in bulk for academic, corporate, or promotional use. eBook versions and licenses are also available for most titles. For more information, reference our Print and eBook Bulk Sales web page at www.apress.com/bulk-sales.

Any source code or other supplementary material referenced by the author in this book is available to readers on GitHub via the book's product page, located at www.apress.com/9781484237205. For more detailed information, please visit www.apress.com/source-code.

Printed on acid-free paper

Table of Contents

About the Author

Timothy Masters earned a PhD in mathematical statistics with a specialization in numerical computing in 1981. Since then he has continuously worked as an independent consultant for government and industry. His early research involved automated feature detection in high-altitude photographs while he developed applications for flood and drought prediction, detection of hidden missile silos, and identification of threatening military vehicles. Later he worked with medical researchers in the development of computer algorithms for distinguishing between benign and malignant cells in needle biopsies. For the past 20 years he has focused primarily on methods for evaluating automated financial market trading systems. He has authored the following books on practical applications of predictive modeling: *Deep Belief Nets in C++ and CUDA C: Volume 2* (Apress, 2018); *Deep Belief Nets in C++ and CUDA C: Volume 1* (Apress, 2018); *Assessing and Improving Prediction and Classification* (Apress, 2018); *Data Mining Algorithms in C++* (Apress, 2018); *Neural, Novel, and Hybrid Algorithms for Time Series Prediction* (Wiley, 1995); *Advanced Algorithms for Neural Networks* (Wiley, 1995); *Signal and Image Processing with Neural Networks* (Wiley, 1994); and *Practical Neural Network Recipes in C++* (Academic Press, 1993).

About the Technical Reviewer

Chinmaya Patnayak is an embedded software developer at NVIDIA and is skilled in C++, CUDA, deep learning, Linux, and filesystems. He has been a speaker and instructor for deep learning at various major technology events across India. Chinmaya earned a master's degree in physics and a bachelor's degree in electrical and electronics engineering from BITS Pilani. He previously worked with the Defense Research and Development Organization (DRDO) on encryption algorithms for video streams. His current interest lies in neural networks for image segmentation and applications in biomedical research and self-driving cars. Find more about him at chinmayapatnayak.github.io.

Introduction

This book is a continuation of Volumes 1 and 2 of this series. Numerous references are made to material in the prior volumes, especially in regard to coding threaded operation and CUDA implementations. For this reason, it is strongly suggested that you be at least somewhat familiar with the material in Volumes 1 and 2. Volume 1 is especially important, as it is there that much of the philosophy behind multithreading and CUDA hardware accommodation appears.

All techniques presented in this book are given modest mathematical justification, including the equations relevant to algorithms. However, it is not necessary for you to understand the mathematics behind these algorithms. Therefore, no mathematical background beyond basic algebra is necessary.

The two main purposes of this book are to present important convolutional net algorithms in thorough detail and to guide programmers in the correct and efficient programming of these algorithms. For implementations that do not use CUDA processing, the language used here is what is sometimes called *enhanced C*, which is basically C that additionally employs some of the most useful aspects of C++ without getting into the full C++ paradigm. Strict C (except for CUDA extensions) is used for the CUDA algorithms. Thus, you should ideally be familiar with C and C++, although my hope is that the algorithms are presented sufficiently clearly that they can be easily implemented in any language.

This book is divided into four chapters. The first chapter reviews feedforward network issues, including the important subject of backpropagation of errors. Then, these issues are expanded to handle the types of layers employed by convolutional nets. This includes locally connected layers, convolutional layers, and several types of pooling layers. All mathematics associated with computing forward-pass activations and backward-pass gradients is covered in depth.

The second chapter presents general-purpose C++ code for implementing the various layer types discussed in the first chapter. Extensive references are made to equations given in the prior chapter so that you are able to easily connect code to mathematics.

The third chapter presents CUDA code for implementing all convolutional net algorithms. Again, there are extensive cross-references to prior theoretical and mathematical discussions so that the function of every piece of code is clear. The chapter ends with a C++ routine for computing the performance criterion and gradient by calling the various CUDA routines.

The last chapter is a user manual for the CONVNET program. This program can be downloaded for free from my web site.

All code shown in the book can be downloaded for free either from my web site (`www.timothymasters.info/deep-learning.html`) or via the Download Source Code button on the book's Apress product page (`www.apress.com/9781484237205`). The complete source code for the CONVNET program is not available, as much of it is related to my vision of the user interface. However, you have access to every bit of code needed for programming the core convolutional net routines. All you need to supply is the user interface.

CHAPTER 1

Feedforward Networks

Convolutional nets are multiple-layer feedforward networks (*MLFNs*) having a special structure that makes them especially useful in computer vision. In this chapter, we will review MLFNs and then show how their structure can be specialized for image processing.

Review of Multiple-Layer Feedforward Networks

A multiple-layer feedforward network is generally illustrated as a stack of layers of "neurons" similar to what is shown in Figure 1-1 and Figure 1-2. The bottom layer is the input to the network, what would be referred to as the *independent variables* or *predictors* in traditional modeling literature. The layer above the input layer is the *first hidden layer*. Each neuron in this layer attains an *activation* that is computed by taking a weighted sum of the inputs, plus a bias, and then applying a nonlinear function. In the fully general case, each hidden neuron in this layer will have a different set of input weights.

If there is a second hidden layer, the activations of each of its neurons is computed by taking a weighted sum of the activations of the first hidden layer, plus a bias, and applying a nonlinear function. This process is repeated for as many hidden layers as desired.

The topmost layer is the output of the network. There are many ways of computing the activations of the output layer, and several of them will be discussed later in the book. For now let's assume that the activation of each output neuron is just a weighted sum of the activations of the neurons in the prior layer, plus a bias, without use of a nonlinear function.

© Timothy Masters 2018
T. Masters, *Deep Belief Nets in C++ and CUDA C: Volume 3*, https://doi.org/10.1007/978-1-4842-3721-2_1

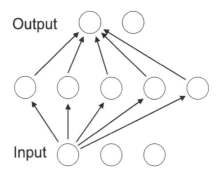

Figure 1-1. *A shallow network*

In Figures 1-1 and 1-2, only a small subset of the connections is shown. Actually, every neuron in every layer feeds into every neuron in the next layer above.

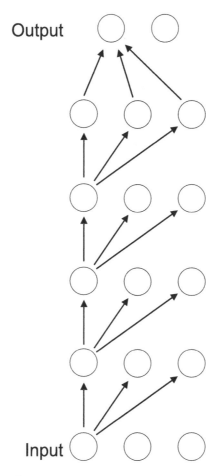

Figure 1-2. *A deep network*

To be more specific, the activation of a hidden neuron, expressed as a function of the activations of the prior layer, is shown in Equation 1-1. In this equation, $x = \{x_1, ..., x_K\}$ is the vector of prior-layer activations, $w = \{w_1, ..., w_K\}$ is the vector of associated weights, and b is a bias term.

$$a = f\left(b + \sum_{k=1}^{K} w_k x_k\right) \tag{1-1}$$

It's often more convenient to consider the activation of an entire layer at once. In Equation 1-2, the weight matrix W has K columns, one for each neuron in the prior layer, and as many rows as there are neurons in the layer being computed. The bias and layer inputs are column vectors. The nonlinear activation function is applied element-wise to the vector.

$$a = f(b + Wx) \tag{1-2}$$

There is one more way of expressing the computation of activations that is most convenient in some situations. The bias vector b can be a nuisance, so it can be absorbed into the weight matrix W by appending it as one more column at the right side. We then augment the x vector by appending 1 to it: $x = \{x_1, ..., x_K, 1\}$. The equation for the layer's activations then simplifies to the activation function operating on a simple matrix/vector multiplication.

$$a = f(Wx) \tag{1-3}$$

What about the activation function? Traditionally, the hyperbolic tangent function has been used because it has some properties that make training faster. This is what we will use here. The hyperbolic tangent function is shown in Equation 1-4 and graphed in Figure 1-3.

$$\tanh(t) = \frac{e^t - e^{-t}}{e^t + e^{-t}} \tag{1-4}$$

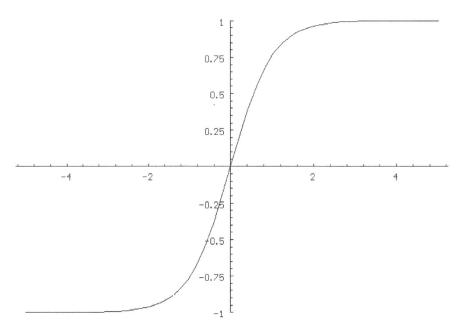

Figure 1-3. *Hyperbolic tangent function*

Wide vs. Deep Nets

Prior to the development of neural networks, researchers generally relied on large doses of human intelligence when designing prediction and classification systems. One would measure variables of interest and then brainstorm ways of massaging these "raw" variables into new variables that (at least in the mind of the researcher) would make it easier for algorithms such as linear discriminant analysis to perform their job. For example, if the raw data were images expressed as arrays of gray-level pixels, one might apply edge detection algorithms or Fourier transforms to the raw image data and feed the results of these intermediate algorithms into a classifier.

The data-analysis world shook when neural networks, especially multiple-layer feedforward networks, came into being. Suddenly we had prediction and classification tools that, compared to earlier methods, relied to a much lesser degree on human-driven preprocessing. It became feasible to simply present an array of gray-level pixels to a neural network and watch it almost miraculously discover salient class features on its own.

For many years, the prevailing wisdom stated that the best architecture for a feedforward neural network was *shallow* and wide. In other words, in addition to the input (often called the *bottom* layer) and the output (often called the *top* layer), the network would have only one, or perhaps two at most, intervening *hidden* layers. This habit was encouraged by several powerful forces. Theorems were proved showing that in very broad classes of problems, one or two hidden layers were sufficient to solve the problem. Also, attempts to train networks with more than two hidden layers almost always met with failure, making the decision of how many layers to use a moot point. According to the theorems of the day, you didn't need deeper networks, and even if you did want more layers, you couldn't train them anyway. So why bother trying?

The fly in the ointment was the fact that the original selling point of neural networks was that they supposedly modeled the workings of the brain. Unfortunately, it is well known that brains are far from shallow in their innermost computational structure (except for those of a few popular media personalities, but we won't go there). And then new theoretical results began appearing that showed that for many important classes of problems, a network composed of numerous narrow layers would be more powerful than a wider, shallower network having the same number of neurons. In effect, although a shallow network might be *sufficient* to solve a problem, it would require enormous width to do so, while a deep network could solve the problem even though it may be very narrow. Deep networks proved enticing though still enormously challenging to implement.

The big breakthrough came in 2006 when Dr. Geoffrey Hinton et al. published the landmark paper "A Fast Learning Algorithm for Deep Belief Nets." The algorithm described in this paper is generally not used for the training of convolutional nets, so we will not pursue it further here; for details, see Volume 1 of this series. Nevertheless, this algorithm is relevant to convolutional nets in that it allowed researchers to discover the enormous power of deep networks. We will see later that convolutional nets, because of their specialized structure, are much easier to train with conventional algorithms than fully general deep networks.

One of the most fascinating properties of deep belief nets, in their general as well as convolutional form, is their remarkable ability to generalize beyond the universe of training examples. This is likely because the output layer, rather than seeing the raw data, is seeing "universal" patterns in the raw data—patterns that due to their universality are likely to reappear in the general population.

A closely related property of deep belief nets is that they are shockingly robust against overfitting. Every beginning statistics student learns the importance of using many more training cases than optimizable parameters. The standard wisdom is that if one uses 100 cases to train a model with 50 optimizable parameters, the resulting model will learn as much about the noise in the training set as it learns about the legitimate patterns and will hence be worthless. But a properly constructed deep network can contain thousands or even millions of optimizable parameters and still avoid overfitting.

Locally Connected Layers

As a general rule, the more optimizable weights we have in a neural network, the more problems we will have. All else being equal, training time goes up exponentially with the number of parameters being optimized. This is a major reason why, before the advent of specialized training algorithms and specialized network architectures, models having more than two hidden layers were practically unknown. Also, the more parameters we optimize, the more likely we are to overfit the model, treating noise in the training data as if it were authentic information.

When the input to the model is an image, it is often reasonable for neurons in a given layer to respond to only neurons in the prior layer that are nearby in the visual field. For example, a neuron in the upper-left corner of the first hidden layer may, by design, be sensitive to only pixels in the upper-left corner of the input image. It may be overkill to cause a neuron in the upper-left corner of the first hidden layer to react to pixels in the opposite corner of the input image.

By implementing this design feature, we tremendously reduce the number of optimizable weights in the model, yet we do not much reduce the total information capture. Even though the neurons in the first hidden layer may each respond to only nearby input neurons, taken as a whole the set of hidden neurons encapsulates information about the entire input image.

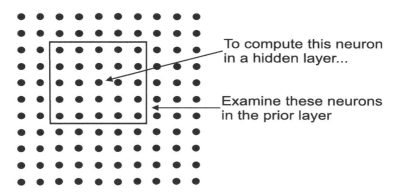

Figure 1-4. *Simple local connections*

Figure 1-4 may be confusing at first. In a conventional neural network, illustrated in Figures 1-1 and 1-2, each layer can be portrayed in one dimension, a line of hidden neurons. But Figure 1-4 has neurons laid out in two dimensions, with its neurons corresponding to those in the prior layer (or input). In fact, it's even more complicated than that. The neural networks presented in this book have three-dimensional layers. Let me explain.

Rows, Columns, and Slices

Think about an input image. It may have multiple bands, such as RGB (red, green, blue). The image has a height (number of rows) and width (number of columns) that are the same for all three bands. In the context of convolutional nets, instead of speaking of bands, we may call them *slices*. In the same way, each hidden layer will occupy a volume described by a height, width, and depth (number of slices). Sometimes the height and width (the *visual field*) of a hidden layer will equal these dimensions of the prior layer, and sometimes they will be less. They will never be greater.

It can be helpful to think of a slice of a hidden layer as corresponding (roughly!) to a single hidden neuron in a conventional neural network. For example, in a conventional network we might have one hidden neuron responding to the sum of two inputs, and a different hidden neuron responding to the difference between these two inputs. In the same way, neurons in one slice may specialize in responding to the total input in the nearby visual field, while neurons in a different slice may specialize in detecting horizontal edges in the nearby visual field. This specialization may vary across the visual field, or it may be forced to be the same across the visual field. We will pursue this concept later.

To compute the activation of a single neuron in a hidden layer, we use an equation similar to Equation 1-1. However, it is considerably more complicated now because it involves only the prior-layer neurons that are nearby in the visual field and all prior-layer slices in this neighborhood. This is roughly expressed in Equation 1-5.

The equation for computing the activation of a single neuron in a locally connected hidden layer involves the following terms:

R : Row of neuron in layer being computed (we call this the *current layer*)

C : Column of neuron in current layer

S : Slice of neuron in current layer

A_{RCS} : Activation of the neuron (or input) being computed

r : Row of neuron in prior layer (or input)

c : Column of neuron in prior layer (or input)

s : Slice of neuron in prior layer (or input)

a_{rcs} : Activation of the prior-layer neuron (or input) at r, c, s

w_{RCSrcs} : Weight associated with the prior-layer neuron (or input) at r, c, s
            when computing the activation of the neuron at RCS

b : The single bias term

$$A_{RCS} = f\left(b + \sum_{s} \sum_{r\,near\,R} \sum_{c\,near\,C} w_{RCSrcs}\, a_{rcs} \right) \tag{1-5}$$

The developer defines what is meant by *near* in the model. Let $NEAR_R$ be the number of prior-layer rows that, by design, are near the row being computed (which we call the *current layer*), and define $NEAR_C$ similarly. Let N_S be the number of slices in the prior layer, the *depth* of that layer. Then the number of weights involved in computing the activation of a neuron is $NEAR_R * NEAR_S * N_S$ plus one for the bias. As a convention in this book, I will often refer to this quantity (including the bias term) as *nPriorWeights*.

Suppose there are N_R rows in the current layer, as well as N_C columns and N_S slices. Then the total number of weights connecting the prior layer to the layer being computed is $N_R * N_C * N_S * nPriorWeights$.

Astute readers will balk at one aspect of this computation. What about the edges of the prior layer, where on one or two sides there are no nearby prior-layer neurons? Great observation! Have patience...we will address this important issue soon.

Convolutional Layers

A few pages ago we mentioned that the pattern in which neurons in a slice specialize may be the same across the visual field, or it may vary. Neither is universally better than the other. If one is dealing with a variety of images, in which specific features do not have a pre-ordained position in the visual field, it probably makes sense for each layer to have a common specialization. For example, all neurons in one slice may respond to the local total brightness, while all neurons in a different slice may contrast the upper part of the local visual field with the lower part and hence be sensitive to a horizontal edge. On the other hand, if the input image is a prepositioned entity, such as a centered face or unknown military vehicle, then it probably makes sense to allow position-relative specialization. For example, neurons a little way in from the top left and top right may specialize in aspects of eye shape on a face.

If the application allows, there is one huge advantage to consistent specialization of a slice across the visual field. In this situation, the weight sets w_{RCSrcs} are the same for all values of R and C, the position in the visual field of the neuron being computed. All neurons across the visual field of a given slice have the same weight set, meaning that the total number of weights connecting the prior layer to the current layer is now just $N_S * nPriorWeights$, which is a lot less than $N_R * N_C * N_S * nPriorWeights$.

Such a layer is called a *convolutional* layer because each of its slices is based on the convolution of the prior layer's activations with the *nPriorWeights* weight set that defines that slice's specialization. (Convolution is a term from filtering theory. If you are unfamiliar with the term, no problem.) For clarity, the activation of a neuron in a convolutional layer is given by Equation 1-6.

$$A_{RCS} = f\left(b + \sum_{s} \sum_{r\,near\,R} \sum_{c\,near\,C} w_{Srcs}\, a_{rcs} \right) \tag{1-6}$$

Half-Width and Padding

So far we have been vague about the meaning of *near* in the visual field. It's time to be specific. Look back at Figure 1-4. We see that in both the vertical and horizontal directions, there are two neurons on either side of the center neuron. This distance is called the *half-width* of the filter. Although the vertical and horizontal half-widths are equal in this example, both being two, they need not be. However, the distance on either side (left-right and up-down from the center) are always equal; otherwise, the center would not be, um, the center. Denote the vertical and horizontal half-widths as HW_V and HW_H, respectively. Then Equation 1-7 gives the number of weights involved in computing the activation of a single neuron. Recall that N_S is the number of slices in the prior layer. The +1 at the end is the bias term.

$$nPriorWeights = N_s \left(2\, HW_H + 1 \right) \left(2\, HW_V + 1 \right) + 1 \tag{1-7}$$

We can now think about edge effect, the problem of a filter extending past the edge of the prior layer into undefined nothingness. We have two extreme options and perhaps a (rarely used) compromise between these two extremes.

1. Instead of letting the leftmost column of the prior layer be the center for the leftmost hidden neuron in the current layer, which causes HW_H columns of needed activation values to be devastatingly undefined, we begin computation HW_H columns inside the left edge. In other words, the leftmost column of the

current layer will have its center in the prior layer at column HW_H instead of the leftmost column. Thus, the intuitively nice alignment will be lost; each column of the current layer will be offset from the corresponding column of the prior layer by HW_H. Similarly, we stop computation HW_H columns before the right edge, and we also inset the top and bottom. This has the advantage of making use of all available information in an exact manner, but it has the disadvantage that rows and columns of the current layer are no longer aligned with rows and columns of the prior layer. This is usually of little or no practical consequence, but it is troubling on a gut level. See Figure 1-5.

2. Pad the prior layer with HW_H columns of zeros on the left and right sides, and HW_V rows of zeros on the top and bottom, to provide "defined" values for the outside-the-visual-field neurons when we place the center of the filter on the edge. This lets us preserve layer-to-layer alignment of neurons in the visual field, which gives most developers a warm, fuzzy feeling and hence is common. It also has an advantage in many CUDA implementations, which I'll touch on in a moment. But it's fraught with danger, as we'll discuss in a moment. See Figure 1-6.

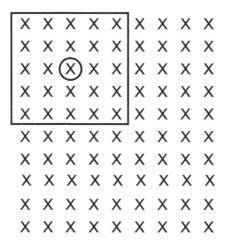

Figure 1-5. *Filter option 1*

0	0	0	0	0	0	0	0	0
0	0	0	0	0	0	0	0	0
0	0	Ⓧ	X	X	X	X	0	0
0	0	X	X	X	X	X	0	0
0	0	X	X	X	X	X	0	0
0	0	X	X	X	X	X	0	0
0	0	X	X	X	X	X	0	0
0	0	X	X	X	X	X	0	0
0	0	0	0	0	0	0	0	0
0	0	0	0	0	0	0	0	0

Figure 1-6. *Filter option 2*

In Figures 1-5 and 1-6, the square box outlines the neurons in the visual field of the prior layer that impact the activation of the top-left neuron in a slice of the current layer. The center of the box is circled. The top-left *X* in these figures is the top-left neuron in the prior layer. Figure 1-5 shows that the top-left neuron in the slice being computed is centered in the visual field two neurons in and two neurons down from the prior layer's top left. In Figure 1-6, we see that the top-left neuron in the slice being computed also corresponds to the top-left neuron in the prior layer because those zeros let the filter extend past the edge.

But make no mistake, those zeros have an impact. It's easy to dismiss them as "nothing" numbers. This feeling is made all the more acceptable because when we program this, we simply avoid adding in the components of Equation 1-5 that correspond to the overhang. Hey, if you don't add them in, they can't do any harm, right? Those weights are just ignored.

Unfortunately, zero is not nothing; it is an honest-to-goodness number. For example, suppose the prior layer is an input image, scaled 0–255. Then zero is pure black! If the weight set computes an average luminance, these zeros will pull the average well down into gray even if the legitimate values are bright. If the weight set detects edges and the legitimate values are bright, a profound edge will be flagged here. For this reason, I am cautious about zero padding. On the other hand, it appears to be more or less standard. You pays your money, and you take your choice.

This fact does, however, provide powerful motivation for using a neuron activation function that is symmetric around zero, such as the hyperbolic tangent shown in Equation 1-4. If one were to use a strictly positive activation function such as the logistic

11

function, the effect of zero padding would be even more severe. Also note that in my CONVNET program, I rescale input images to minus one through one rather than the more common 0–255. This lessens the impact of zero padding.

I should add that full zero padding can be advantageous in many CUDA implementations. This will be discussed in detail later when we explore CUDA code, but the idea is that certain numbers of hidden neurons, such as multiples of 32, speed operation by making memory accesses more efficient. On the other hand, lack of full zero padding impacts only the size of the visual field, not the depth, and good CUDA implementations can compensate for shrinking visual fields by handling the depth dimension properly.

Note that one is not bound to employ one of these two extreme options. It is perfectly legitimate to compromise and pad with fewer than HW_H columns of zeros on the left and right, and HW_V rows of zeros on the top and bottom. Nobody seems to do it, but you needn't let that stop you.

Striding and a Useful Formula

A common general principle of neural network design is that the size of hidden layers decreases as you move from input toward output. Of course, we can (and usually do) decrease the depth (number of slices) of successive layers. But effective information compression is also obtained by decreasing the size of the visual field (rows and columns) in successive layers. If we pad with half-width zeros as in option 2 in the prior section, the size of the visual field remains constant. And even if we do not pad, the visual field only slightly decreases. There is a more direct approach: striding.

It should be emphasized that the modern tendency is to avoid striding and use *pooling* to reduce the visual field. That topic will be discussed later in the chapter. However, because striding does have a place in our toolbox, we'll cover it now.

The idea of striding is simple: instead of marching the centers of the prior layer and the current layer together, moving each one place at a time, we move the prior layer neurons faster. For example, we might move the prior layer twice as fast as the current layer. Suppose we have fully padded so that row 1, column 1 in the current layer is centered on row 1, column 1 of the prior layer. Then row 1, column 2 of the current layer is centered at row 1, column 3 of the prior layer, and so forth. Each time we move one row/column in the current layer, we move two rows/columns in the prior layer. This cuts the number of rows/columns approximately in half (or whatever the stride factor is), hence reducing the number of neurons in the visual field by a factor of the square of the striding value.

We now present a simple formula for the number of rows/columns in the current layer, given the size of the prior layer and the size of the filter, the amount of zero padding, and the stride. No identification of vertical or horizontal is needed, as this formula applies to each dimension. The following definitions for the terms of the formula in Equation 1-8 apply:

W: Width/height of the prior layer
F: Width/height of the filter; two times half-width, plus one
P: Padding rows/columns appended to each edge; less than or equal to half-width
S: Stride
C: Width/height of the current layer

$$C = (W - F + 2P)/S + 1 \tag{1-8}$$

There is widespread belief that the division by the stride must be exact; if the numerator is not a multiple of the stride, the layer is somehow invalid. A brief Internet search shows this belief to be ubiquitous. But it's not really true. There are two things that make this belief appealing.

- If the division is not exact, the alignment of the current layer with the prior layer will not be symmetric; the current layer may be inset from the prior layer by different amounts on the right and left, or top and bottom. However, I do not see any reason in any application why this lack of symmetry would be a problem. If this is a problem in your application, then select your parameters in such a way as to make the division exact. But it's silly for the padding to exceed the half-width, and the filter size may be important and not amenable to change. This can make it difficult to produce perfect division.

- Many popular training algorithms, which generally use packaged matrix multiplication routines, require exact division. So if you use such an algorithm, you have no choice. The algorithms presented in this book and employed in the CONVNET program do not impose this requirement.

Pooling Layers

The prior section discussed striding, a means of reducing the size of the visual field when progressing from one layer to the next. Although this method was popular for some time and is still occasionally useful, it has recently been supplanted by the use of a *pooling* layer. In particular, the stride of a locally connected or convolutional layer is generally kept at one so that the visual field is left unchanged (if full padding) or only slightly reduced (if less than full padding). Then, a layer whose sole purpose is to reduce the visual field is employed.

Pooling layers are similar to locally connected/convolutional layers in that they move a rectangular window across the prior layer, applying a function to the activation values in each window to compute the activation of a single neuron in the current layer. But the biggest difference is that pooling layers are not trainable. Their function, which maps window values in the prior layer to an activation in the current layer, is fixed in advance.

There are three other differences. Padding is generally not used; it is avoided in this book, as I believe the distortion introduced by padding a pooling layer is too risky. Also, filter widths can be even; they do not take the form 2*HalfWidth+1. The implication is that pooling destroys layer-to-layer alignment.

Finally, the pooling function that maps the prior layer to the current layer is applied separately to each slice. The locally connected/convolutional layers discussed in the previous few sections look at all prior-layer slices simultaneously. So, for example, if we have a five-by-five filter operating on a prior layer that has ten slices, a total of 5*5*10=250 activations in the prior layer take part in computing the activation of a neuron in the current layer. But in a pooling layer, there are as many slices as in the prior layer, and each layer is computed independently. So, using these same numbers, each of the ten neurons in the current layer occupying the same position in the visual field would be computed from 25 prior-layer activations in the corresponding slice. We map first slice to first slice, second slice to second slice, and so forth.

Pooling Types

Historically, the first type of pooling was *average pooling*. The mapping function simply takes the average of the activations in the window placed on the prior layer. Average pooling has recently fallen out of favor, but some developers still find it appropriate in some applications.

The most popular type of pooling as of this writing is *max pooling*. This mapping function chooses the neuron in the prior layer's window, which has maximum activation. Much experience indicates that this is more effective than average pooling.

One small but annoying disadvantage of max pooling is that it is not differentiable everywhere. At the activation levels where the choice transitions from one neuron to another, the derivative of the performance criterion with respect to a particular weight goes to zero on the neuron suddenly losing the contest and jumps away from zero on the winner. This slightly impedes some optimization algorithms, and it makes numerical verification of gradient computations a bit dicey. But in practice, these problems do not seem to be overly serious, so we put up with them.

Other pooling functions are appearing. Different norms can be used, and some even more exotic functions have been proposed. None of these alternatives is discussed in this book.

The Output Layer

This book, as well as the CONVNET program, follows the simple convention that the output layer contains one neuron for each class. Each of these neurons is fully connected to all neurons in the prior layer. Because the concept of visual field makes no sense in the concept of output-layer classes, this layer by definition is organized as a single row and column (the "visual field" is one pixel) with a depth (number of slices) equal to the number of classes. The exact organizational layout is not vital, but this layout proves to simplify programming and mathematical derivations.

SoftMax Outputs

Back in the olden days when I was a graduate student, classification performed with numerical prediction models was typically done by having as many predicted outputs as there are classes and assigning a target value of 1.0 to the output corresponding to the correct class and of 0.0 for all of the incorrect classes. When the model was put to use, whichever output had the largest prediction was chosen as the predicted class. The exact values of the predictions usually had little theoretical or practical meaning; we just picked the largest. One might call this a "hard" selection process.

In these more enlightened times, we can "soften" the selection process, making the predicted outputs resemble probabilities. This is extremely useful, not just because it's nice to be able to talk about the predicted probability of each class (even though in many applications this interpretation is excessively optimistic!) but also for an even more important reason. These *SoftMax* outputs make the model far more robust against outliers in the training and test data. This vital topic is discussed in detail in Volume 1 of this series, so it will be glossed over here. But we do need to review the relevant equations that we will program.

We know that the activation of a single hidden neuron is computed as a nonlinear function of a weighted average of prior-layer activations (plus a bias term). For the output neurons we drop the nonlinear function and speak only of the weighted average (plus bias). This quantity is called the *logit* of the neuron being computed. This is shown in Equation 1-9 for output neuron k. In this equation, $\boldsymbol{x} = \{x_1, x_2, ...\}$ is the vector of activations of the final hidden layer, $\boldsymbol{w} = \{w_{k1}, w_{k2}, ...\}$ is the vector of associated weights, and b_k is a bias term. In other words, the logit of an output neuron is computed exactly like we compute the activation of a hidden-layer neuron, except that we do not apply the nonlinear activation function.

$$\boldsymbol{logit}_k = \boldsymbol{b}_k + \sum_i \boldsymbol{w}_{ki} \, \boldsymbol{x}_i \tag{1-9}$$

Once we have the logit of every output neuron, computing the SoftMax output values, which can roughly be thought of as probabilities of class membership, is done with Equation 1-10. This equation assumes that there are K output neurons (classes). It should be obvious that these output activations are non-negative and sum to one.

$$\boldsymbol{p}(\boldsymbol{y} = \boldsymbol{k}) = \frac{e^{logit_k}}{\sum_{i=1}^{K} e^{logit_i}} \tag{1-10}$$

The traditional mean squared error optimization criterion is of little value when dealing with SoftMax outputs. We now need a different optimization criterion to find good values for the parameters of the model. An excellent choice is *maximum likelihood*. This is not the venue for a detailed description of maximum likelihood, but we will try for an intuitive justification.

Any set of model parameters defines, by means of the equations just shown, the probability of each possible class given an observed case. Our training set is assumed to be random draws from a population, each of which provides an input vector and a true class. If we were to consider a given set of model parameters as defining the true model, we could compute (in a sense best left undiscussed here) the probability of obtaining the set of training cases that were actually observed. So we find that set of parameters that maximizes this probability. In other words, we seek the model that provides the maximum likelihood of having obtained our training set in these random draws from the population.

In our particular application, the likelihood of a case is just the probability given by the model for the class to which that case belongs. We want a criterion that is summable across the training set, so instead of considering the likelihood, which is multiplicative, we will use the log likelihood as our criterion. This way we can compute the criterion for the entire training set by summing the values for the individual cases in the training set.

Also, to conform to more general forms of the log likelihood function that you may encounter in more advanced texts, as well as to conform to the expression of the derivative that will soon be discussed, we express the log likelihood of a case in a more complex manner. For a given training case, define t_k as 1.0 if this case is a member of class k, and 0.0 otherwise. Also define p_k as the SoftMax activation of output neuron k, as given by Equation 1-10. Then, for our single training case, the log of the likelihood corresponding to the model's parameters is given by Equation 1-11. This equation is called the *cross entropy*, and interested readers might want to look up this term for some fascinating insights.

$$L = \sum_{k=1}^{K} t_k \log(p_k) \qquad (1\text{-}11)$$

Observe that in the summation over classes, every term is zero except the term corresponding to the correct class. Thus, the log likelihood is just the log of the model's computed probability for the correct class of the case. Here are some observations about the log likelihood:

- Because p is less than one, the log likelihood is always negative.

- The better the model is at computing the correct class probabilities, the larger (closer to zero) this quantity will be since it is the log probability of the correct class and a good model will provide a large probability for the correct class.

- If the model is nearly perfect, meaning that the computed probability of the correct class is nearly 1.0 for every case, the log likelihood will approach zero, its maximum possible value.

We will soon discuss gradient computation, at which time we will need the derivative of the log likelihood. Without going through the considerable number of steps, we state that this derivative of Equation 1-11 for a case is given by Equation 1-12.

$$\delta_k^o = \frac{\partial L}{\partial logit_k} = p_k - t_k \tag{1-12}$$

Developers with experience in computing the gradient of traditional neural networks will be amazed to see that, except for a factor of two, the delta for a SoftMax output layer and maximum likelihood optimization is identical to that for a linear output layer and mean-squared-error optimization. This means that traditional predictive model gradient algorithms can be used for SoftMax classification with only trivial modification. Nonetheless, we will summarize gradient computation in the next section.

Back Propagation of Errors for the Gradient

The fundamental goal of supervised training can be summarized simply: find a set of parameters (weights and biases as in Equation 1-2) such that, given an input to the neural network, the output of the network is as close as possible to the desired output. To find such parameters, we must have a performance criterion that rigorously defines the concept of "close." We then find parameters that optimize this criterion.

Suppose we have K output neurons numbered 1 through K. For a given training case, let t_k be the true value for this case, the value that we hope the network will produce, and let p_k be the output actually obtained. Then the log likelihood for this single case is given by Equation 1-11. To compute the log likelihood for the entire training set, sum this quantity for all cases. To keep this quantity to "reasonable" values, most people (including me) divide this sum by the number of cases and the number of classes. If there are N training cases, this performance criterion is given by Equation 1-13.

$$L_{tset} = \frac{\sum_{i=1}^{N} L_i}{KN} \tag{1-13}$$

Supervised training of a multiple-layer feedforward network amounts to finding the weights and bias terms that maximize Equation 1-13 (or minimize its negative, which is what we really do). In any numerical minimization algorithm, it is of great benefit to be able to efficiently compute the gradient, the partial derivatives of the criterion being minimized with respect to each individual parameter. Luckily, this is quite easy in this application. We just start at the output layer and work backward, repeatedly invoking the chain rule of differentiation.

The activation of output neuron k is given by Equation 1-10. Neural net aficionados use the Greek letter delta to designate the derivative of the performance criterion with respect to the net input coming into a neuron; in the current context this is output neuron k, and its delta is given by Equation 1-12.

In other words, this neuron is receiving a weighted sum of activations from all neurons in the prior layer, and from Equation 1-12 we know the derivative of the log likelihood criterion with respect to this weighted sum.

How can we compute the derivative of the criterion with respect to the weight from neuron i in the prior layer? The simple chain rule tells us that this is the product of the derivative in Equation 1-12 times the derivative of the net input (the weighted sum coming into this output neuron) with respect to this weight.

This latter term is trivial. The contribution to the weighted sum from neuron i in the prior layer is just the activation of that neuron times the weight connecting it to the output neuron k. We shall designate this output weight as $w_{ki}{}^O$. So the derivative of that weighted sum with respect to $w_{ki}{}^O$ is just the activation of neuron i. This leads us to the formula for the partial derivatives of the criterion with respect to the weights connecting the last hidden layer to the output layer. In Equation 1-14 we use the superscript M on a to indicate that it is the activation of a neuron in hidden layer M, where there are M hidden layers numbered from 1 through M.

$$\frac{\partial \boldsymbol{L}}{\partial \boldsymbol{w}_{ki}^{O}} = \boldsymbol{a}_i^M \delta_k^o \tag{1-14}$$

There are two complications when we deal with the weights feeding hidden layers. Let's consider the weights leading from hidden layer $M-1$ to hidden layer M, the last hidden layer. We ultimately want the partial derivatives of the criterion with respect to each of these weights. As when dealing with the output layer, we'll split this derivative into the product of the derivative of the net input feeding this neuron with respect to the weight times the derivative of the criterion with respect to this neuron's net input.

As before, the former term here is trivial: just the activation of the prior neuron feeding through this weight. It's the latter that's messy.

The first complication is that the hidden neurons are nonlinear. In particular, the function that maps the net input of a hidden neuron to its activation is the hyperbolic tangent function shown in Equation 1-4. So the chain rule tells us that the derivative of the criterion with respect to the net input is the derivative of the criterion with respect to the output times the derivative of the output with respect to the input. Luckily, the derivative of the hyperbolic tangent function $f(a)$ is simple, as shown in Equation 1-15.

$$f'(a) = 1 - f^2(a) \qquad (1\text{-}15)$$

The remaining term is more complicated because the output of a neuron in a hidden layer feeds into every neuron in the next layer and thus impacts the criterion through every one of those paths. Recall that $\delta_k{}^o$ is the derivative of the criterion with respect to the weighted sum coming into output neuron k. The contribution to this weighted sum going into output neuron k from neuron i in the prior layer M is the activation of hidden neuron i times the weight connecting it to output neuron k. So the impact on the derivative of the criterion from the activation of neuron i that *goes through this path* is $\delta_k{}^o$ times the connecting weight. Since neuron i impacts the error through all output neurons, we must sum these contributions, as shown in Equation 1-16.

$$\frac{\partial L}{\partial a_i^M} = \sum_{k=1}^{K} w_{ki}^o \, \delta_k^o \qquad (1\text{-}16)$$

Pant pant. We are almost there. Our goal, the partial derivative of the criterion with respect to the weight connecting a neuron in hidden layer $M{-}1$ to a neuron in hidden layer M is the product of the three terms that we have already presented.

- The derivative of the net input to the neuron in hidden layer M with respect to the weight in which we are interested

- The derivative of the output of this neuron with respect to its net input (the derivative of its nonlinear activation function)

- The derivative of the criterion with respect to the output of this neuron

The derivative of the criterion with respect to $w_{ij}{}^M$ (the weight connecting neuron j in layer $M{-}1$ to neuron i in layer M) is the product of these three terms. The product of the second and third of these terms is given by Equation 1-17, with $f'(.)$ being given by Equation 1-15. The multiplication is completed in Equation 1-18.

$$\delta_i^M = f'\left(a_i^M\right) \sum_{k=1}^{K} w_{ki}^O\, \delta_k^O \tag{1-17}$$

$$\frac{\partial L}{\partial w_{ij}^M} = a_j^{M-1} \delta_i^M \tag{1-18}$$

There is no need to derive the equations for partial derivatives of weights in hidden layers prior to the last hidden layer, as the equations are the same, just pushed back one layer at a time by successive application of the chain rule. In particular, for some hidden layer $m<M$, we have Equation 1-19 for the partial derivative of the criterion with respect to the weighted sum coming into neuron i in layer m. Equation 1-20 then provides the partial derivative of the criterion with respect to the weight connecting neuron j in hidden layer $m{-}1$ to neuron i in hidden layer m. In this case, there are K neurons in hidden layer $m{+}1$.

$$\delta_i^m = f'\left(a_i^m\right) \sum_{k=1}^{K} w_{ki}^{m+1} \delta_k^{m+1} \tag{1-19}$$

$$\frac{\partial E}{\partial w_{ij}^m} = a_j^{m-1} \delta_i^m \tag{1-20}$$

That was a long haul, especially for those for whom math is not pleasant. So as an aid to those who are mainly interested in programming, here is a more concise summary of the procedure for computing the gradient:

1. Allocate two scratch vectors, this_delta[] and prior_delta[]. These must be as long as the maximum number of hidden neurons in any layer, as well as the number of classes (output neurons).

2. Compute activations for all hidden layers and the output layer.

3. Use Equation 1-12 to compute the output deltas. Put these in this_delta.

4. Use Equation 1-14 to compute the gradient of the output layer.

5. Designate the last hidden layer as the "current" layer, which makes the output layer the "next" layer.

6. This is the beginning of the main loop that moves backward through the network, from the last hidden layer to the first. At this time, this_delta[k] contains the derivative of the criterion with respect to the input (post-weight) to neuron k in the next layer.

7. Backpropagate delta. To get the contribution of that neuron k from neuron i in the current layer, the layer whose gradient is currently being computed, we multiply delta[k] by the weight connecting current-layer neuron i to next-layer neuron k. This gives us the part of the total derivative due to the output of neuron i in the current layer going through neuron k in the next layer. But the output of neuron i impacts the criterion derivative through *all* neurons in the next layer. Thus, we must sum these parts across all neurons (values of k) in the next layer. To get the derivative of the criterion with respect to the input to neuron i, we multiply this sum by the derivative of neuron i's activation function. This is Equation 1-19, or Equation 1-17 if this is the last hidden layer. The arguments for this equation are in this_delta, and we put the results in prior_delta.

8. Move the contents of prior_delta to this_delta.

9. To get the derivative of the criterion with respect to a weight coming into neuron i, we multiply delta by the input coming through this weight (the output of the prior layer's neuron). This is Equation 1-20, or Equation 1-18 if this is the last hidden layer. If there are more hidden layers to process, go to step 6.

Even though we will be dealing with specialized types of layers, such as locally connected, convolutional, and pooling layers, the steps just described apply for all. We merely have to be careful to identify items that are identically zero and hence ignored. In the conventional implementation (page 42), we get the deltas for step 9 from prior_delta, so we can perform step 8 after step 9 is complete. In the CUDA version (page 111), we will get the deltas for step 9 from this_delta, so we must perform step 8 before step 9.

CHAPTER 2

Programming Algorithms

The source code that can be downloaded for free from my web site contains four large source files that handle the vast majority of the computation involved in propagating activations and backpropagating deltas for all layer types involved in convolutional nets.

- **MOD_NO_THR.CPP**: Nonthreaded versions of all routines. These are not used in the CONVNET program, but they are the routines listed and discussed in this book. Because they are not designed for threaded use, they are somewhat simpler than the threaded versions. In this way, the focus of discussion can be on the algorithms themselves, avoiding the complexities of threading.

- **MOD_THR.CPP**: Threaded versions of all routines. The last section of this chapter will explore how they differ from the nonthreaded versions and how they are incorporated into a fully multithreaded program.

- **MOD_CUDA.CPP**: Host routines that call the CUDA routines and coordinate all CUDA-based computation.

- **MOD_CUDA.cu**: All CUDA source code, as well as their host-code wrappers. Note that *cu* is lowercase. For some bizarre reason, Visual Studio has problems when it is in uppercase. Go figure.

Here is the order in which routines will be presented in this chapter:

1. Extract of Model declaration, showing key declarations

2. Extract of Model constructor, showing how architecture is built

3. trial_no_thr(), externally callable routine that computes all activations

4. Activation functions for each layer type; called from trial_no_thr()

23

© Timothy Masters 2018
T. Masters, *Deep Belief Nets in C++ and CUDA C: Volume 3*, https://doi.org/10.1007/978-1-4842-3721-2_2

5. trial_error_no_thr(), externally callable routine to compute criterion

6. grad_no_thr(), externally callable routine to compute gradient

7. Gradient routines for each layer type; called from grad_no_thr()

8. Backprop routines for each layer type; called from gradient routines

9. Discussion of threading the algorithms

Model Declarations

The complete set of model declarations can be found in the file CLASSES.H. However, most of them are irrelevant to the discussion of the activation and gradient algorithms, so they are not printed in the text.

Also, there are a handful of variables used so extensively that I (please forgive me!) made them global. They are as follows:

```
int n_pred;              // Number of predictors present (input rows*cols*bands)
int n_classes;           // Number of classes
int n_db_cols;           // Size of a case in the database = n_pred + n_classes
int n_cases;             // Number of cases (rows) in database
double *database;        // They are here, variables changing fastest

int IMAGE_rows;          // Input number of rows
int IMAGE_cols;          // and columns
int IMAGE_bands;         // Its number of bands
```

Here are the important Model class declarations for convenient reference. Note that some duplicate globals. The declarations that are arrays have separate values for each layer.

```
int n_pred;              // Number of predictors present (input grid size; rows*cols*bands)
int n_classes;           // Number of classes
int n_layers;            // Number of hidden layers (does not include input or output)
int layer_type[];        // Each entry is type of layer
int height[];            // Number of neurons vertically in a slice of this layer
int width[];             // Ditto horizontal; these are both 1 for a fully connected layer
int depth[];             // Number of slices in this layer; number of hidden if fully connected
```

```
int nhid[];                    // Number of neurons in this layer = height times width times depth
int HalfWidH[];                // Horizontal half width looking back to prior layer
int HalfWidV[];                // And vertical
int padH[];                    // Horizontal padding, must not exceed half width
int padV[];                    // And vertical
int strideH[];                 // Horizontal stride
int strideV[];                 // And vertical
int PoolWidH[];                // Horizontal pooling width looking back to prior layer
int PoolWidV[];                // And vertical
int n_prior_weights[];         // N of inputs per neuron (including bias) from prior layer
                               // = prior depth * (2*HalfWidH+1) * (2*HalfWidV+1) + 1
                               // A CONV layer has this many weights per slice
                               // A LOCAL layer has this times its nhid
int n_hid_weights;             // Total number of all hidden weights; includes bias
int n_all_weights;             // As above, but also includes output layer weights
int max_any_layer;             // Max n of neurons in any layer, including input and output
double *weights;               // All 'n_all_weights' weights, including final weights, are here
double *layer_weights[];        // Pointers to each layer's weights in 'weight' vector
double *gradient;              // 'n_all_weights' gradient, aligned with weights
double *layer_gradient[];       // Pointers to each layer's gradient in 'gradient' vector
double *activity[];            // Activity vector for each layer
double *this_delta;            // Scratch vector for gradient computation
double *prior_delta;           // Ditto
double output[];               // SoftMax activation for each class
int *poolmax_id[];             // Used only for POOLMAX layer; saves from forward pass ID
```

Order of Weights and Gradient

The weights for layer *i* begin at layer_weights[i]. Similarly, the gradient (which aligns element by element with the corresponding weights) for layer *i* begin at layer_gradient[i].

Two general ordering rules govern all layer types.

1. Within each layer the weights (and gradient) are ordered with the input to the layer changing faster than the neuron being computed.

2. The width changes fastest, then the height, and finally the depth slowest.

For a fully connected layer, these two rules clearly describe the situation. First we have the n_prior_weights weights connecting the prior layer to the first hidden neuron, with the bias last. Within that vector, the prior layer's width changes fastest, then the height, and finally the depth slowest. After this, we have a similar vector for the second neuron in the current layer, and so forth. Recall that in a fully connected layer, the height and width are both one, with neurons strung out along the depth.

For other layer types, the order is slightly more complex and will be described as each activation routine is presented.

Initializations in the Model Constructor

Most of the code in the Model constructor is mundane and not worth listing in this text. You can see the full module in MODEL.CPP. However, some of this code reinforces discussions in the prior chapter and so is presented here.

In the loop shown next, we compute n_prior_weights in three steps for locally connected and convolutional layers. First we set it equal to the size of the moving-window filter, the number of weights in the filter. Then we multiply this by the number of slices in the prior layer because the filter is applied to all prior-layer slices simultaneously. Finally, we add 1 to include the bias term. Also in this loop we use Equation 1-8 to compute the size of the visual field.

```
for (i=0; i<n_layers; i++) {
  nfH = 2 * HalfWidH[i] + 1;        // Filter width
  nfV = 2 * HalfWidV[i] + 1;

  if (layer_type[i] == TYPE_LOCAL || layer_type[i] == TYPE_CONV) {
    n_prior_weights[i] = nfH * nfV; // Inputs, soon including bias, to neurons in layer
    if (i == 0) {
      height[i] = (IMAGE_rows - nfV + 2 * padV[i]) / strideV[i] + 1;
      width[i] = (IMAGE_cols - nfH + 2 * padH[i]) / strideH[i] + 1;
      n_prior_weights[i] *= IMAGE_bands;
      }
    else {
      height[i] = (height[i-1] - nfV + 2 * padV[i]) / strideV[i] + 1;
      width[i] = (width[i-1] - nfH + 2 * padH[i]) / strideH[i] + 1;
      n_prior_weights[i] *= depth[i-1];
      }
    n_prior_weights[i] += 1;         // Include bias
    }
```

By common convention, a fully connected layer is implemented as a one-pixel visual field, with a slice for each neuron. It has a weight from every prior-layer activation, plus the bias term.

```
else if (layer_type[i] == TYPE_FC) {
  height[i] = width[i] = 1;
  if (i == 0)
    n_prior_weights[i] = n_pred + 1;
  else
    n_prior_weights[i] = nhid[i-1] + 1;
  }
```

Pooling layers also have their visual field size defined by Equation 1-8. They align slice by slice with the prior layer, each processed independently, so a pooling layer has the same number of slices as the prior layer. Padding is never used (by me anyway) for pooling layers. Pooling layers are a fixed function, with no trainable weights, so n_prior_weights is zero. Finally, the number of hidden neurons in this layer, regardless of type, is the product of the dimensions.

```
else if (layer_type[i] == TYPE_POOLAVG || layer_type[i] == TYPE_POOLMAX) {
  if (i == 0) {
    height[i] = (IMAGE_rows - PoolWidV[i]) / strideV[i] + 1;
    width[i] = (IMAGE_cols - PoolWidH[i]) / strideH[i] + 1;
    depth[i] = IMAGE_bands;
    }
  else {
    height[i] = (height[i-1] - PoolWidV[i]) / strideV[i] + 1;
    width[i] = (width[i-1] - PoolWidH[i]) / strideH[i] + 1;
    depth[i] = depth[i-1];
    }
  n_prior_weights[i] = 0;
  }

nhid[i] = height[i] * width[i] * depth[i];
}
```

The previous code handles the hidden layers. We do the output layer, which is always fully connected, in the following code. We don't need to worry about the height, width, and depth because they will never be referenced in subsequent code that processes the output layer.

```
if (n_layers == 0)
  n_prior_weights[n_layers] = n_pred + 1;    // Output layer, always fully connected
else
  n_prior_weights[n_layers] = nhid[n_layers-1] + 1;
```

Lastly, we compute the total number of weights in all hidden layers, not including the output layer. We also need the maximum size of any layer, input, hidden, or output. These will be used for memory allocation, not shown here. This code is presented only to reinforce architectural issues in the model.

The most important fact here is that locally connected and fully connected layers have a number of weights equal to n_prior_weights times the number of hidden neurons in the layer because each hidden neuron has its own set of weights. But a convolutional layer has a number of weights equal to n_prior_weights times the depth of this layer because every neuron in the visual field of a given slice shares the same set of weights.

```
max_any_layer = n_pred;          // Input layer is included in max
if (n_classes > max_any_layer)
  max_any_layer = n_classes;     // Output layer is included in max

n_hid_weights = 0;
for (ilayer=0; ilayer<n_layers; ilayer++) { // For each of the hidden layers
  if (nhid[ilayer] > max_any_layer)
    max_any_layer = nhid[ilayer];
  if (layer_type[ilayer] == TYPE_FC || layer_type[ilayer] == TYPE_LOCAL)
    n_hid_weights += nhid[ilayer] * n_prior_weights[ilayer];
  else if (layer_type[ilayer] == TYPE_CONV)
    n_hid_weights += depth[ilayer] * n_prior_weights[ilayer];
  else if (layer_type[i] == TYPE_POOLAVG || layer_type[i] == TYPE_POOLMAX)
    n_hid_weights += 0;          // Just for clarity; pooling has no trainable weights
  } // For ilayer (each hidden layer)

n_all_weights = n_hid_weights + n_classes * n_prior_weights[n_layers]; // Add output
```

Finding All Activations

The routine trial_no_thr() can be called from elsewhere. It does a forward pass to compute all activations in the model. None of the nitty-gritty calculations appears here; the routine simply calls the appropriate specialist for each layer.

```
void Model::trial_no_thr (double *input)
{
  int i, ilayer;
  double sum;

  for (ilayer=0; ilayer<n_layers; ilayer++) {   // These do not include output layer
    if (layer_type[ilayer] == TYPE_LOCAL)
      activity_local_no_thr (ilayer, input);
    else if (layer_type[ilayer] == TYPE_CONV)
      activity_conv_no_thr (ilayer, input);
    else if (layer_type[ilayer] == TYPE_FC)
      activity_fc_no_thr (ilayer, input, 1);
    else if (layer_type[ilayer] == TYPE_POOLAVG ||
           layer_type[ilayer] == TYPE_POOLMAX)
      activity_pool_no_thr (ilayer, input);
    }

  activity_fc_no_thr (n_layers, input, 0);      // Output layer

  // Classifier is always SoftMax. Use Equation 1-10 on Page 16.
  sum = 1.e-60;                         // Denominator below must never be zero
  for (i=0; i<n_classes; i++) {
    if (output[i] < 300.0)             // Be safe against rare but deadly problem
      output[i] = exp (output[i]);
    else
      output[i] = exp (300.0);
    sum += output[i];
    }
  for (i=0; i<n_classes; i++)
    output[i] /= sum;
}
```

Activating a Fully Connected Layer

Computing the activation of a fully connected layer is relatively easy because every neuron in the layer is connected to every neuron in the prior layer. We do not have to worry about the position of a moving window or whether we are past the edge of the prior layer, or striding, and so forth. These considerations can be surprisingly complicated to implement efficiently. Thus, we begin with this easy routine.

One potential source of confusion is the input parameter. This is not the input to the layer being computed; if this layer is past the first hidden layer, the input to this layer will be fetched directly from the activity vector of the prior hidden layer. Rather, this is the input to the model, and it is used only if this is the first layer after the input.

```
void Model::activity_fc_no_thr (int ilayer, double *input, int nonlin)
{
  int iin, iout, nin, nout;
  double sum, *wtptr, *inptr, *outptr;

  wtptr = layer_weights[ilayer];          // Weights for this layer

  if (ilayer == 0) {                      // The 'prior layer' is the input vector
    nin = n_pred;                         // This many elements in the vector
    inptr = input;                        // They are here
    }
  else {                                  // The prior layer is a hidden layer
    nin = nhid[ilayer-1];                 // It has this many neurons
    inptr = activity[ilayer-1];           // Prior layer's activations
    }

  if (ilayer == n_layers) {               // If this is the output layer
    nout = n_classes;                     // There is one output neuron for each class
    outptr = output;                      // Outputs go here
    }
  else {                                  // This is a hidden layer
    nout = nhid[ilayer];                  // We must compute this many activations
    outptr = activity[ilayer];            // And put them here
    }
```

```
for (iout=0; iout<nout; iout++) {        // Compute each activation

    sum = 0.0;
    for (iin=0; iin<nin; iin++)          // Equation 1-1 on Page 3
        sum += inptr[iin] * *wtptr++;
    sum += *wtptr++;                     // Bias

    if (nonlin) {                        // Hidden layers are nonlinear; output is not
        sum = exp (2.0 * sum);           // Hyperbolic tangent function
        sum = (sum - 1.0) / (sum + 1.0); // Equation 1-4 on Page 3
        }
    outptr[iout] = sum;
    }
}
```

Activating a Locally Connected Layer

First, we must be clear on how the weights that connect the prior layer to this locally connected layer are ordered. They can best be visualized as they would be processed in a set of nested loops:

Current layer depth
  Current layer height
   Current layer width
     Prior layer depth
       Prior layer height
         Prior layer width
     Bias

The depth dimension of the neuron being computed changes slowest, then the height, and finally the width. At the width point (three levels in), we are looking at the weights for computing a single neuron in this layer. We have the weights that connect it to the prior layer, in the order shown. After these prior-layer weights appear, we have the single bias term.

The input parameter is the input to the entire model, which will be used only if the layer we are about to compute is the first layer after the input.

```
void Model::activity_local_no_thr (int ilayer, double *input)
{
    int k, in_row, in_rows, in_col, in_cols, in_slice, in_slices, iheight, iwidth, idepth;
    int rstart, rstop, cstart, cstop;
    double sum, *wtptr, *inptr, *outptr, x;

    if (ilayer == 0) {                   // This is the first layer after the input
      in_rows = IMAGE_rows;
      in_cols = IMAGE_cols;
      in_slices = IMAGE_bands;
      inptr = input;                     // Input to this layer is the model's input image
      }

    else {                               // The prior layer is a hidden layer
      in_rows = height[ilayer-1];
      in_cols = width[ilayer-1];
      in_slices = depth[ilayer-1];
      inptr = activity[ilayer-1];        // Input to this layer is the prior layer's activations
      }

    wtptr = layer_weights[ilayer];       // Weights for this layer, order as described above
    outptr = activity[ilayer];           // We put the computed activations here

    k = 0;                               // This will index the computed activations in outptr
    for (idepth=0; idepth<depth[ilayer]; idepth++) {
      for (iheight=0; iheight<height[ilayer]; iheight++) {
        for (iwidth=0; iwidth<width[ilayer]; iwidth++) {

          // Compute activation of this layer's neuron at (idepth, iheight, iwidth)
```

Here's where things get a little complicated. We are about to compute the activation of this neuron. This computation is based on a rectangle in the *prior* layer whose position is determined by the position (iheight, iwidth) of the current neuron in the visual field of *this* layer. In both the vertical and horizontal directions, the center of the first filter (first row or column of the current layer) is at the location *HalfWidth-Pad* in the prior layer, and the first row/column of this first rectangle is at *-Pad,* which will be in the zero-padding area if padding is done. If this is not clear, please draw yourself a little one-dimensional picture using two rows of dots, a row for each layer. Understanding this is crucial!

We can now compute the inclusive starting and stopping rows and columns of the rectangle in the prior layer, which contributes to the activation of the neuron in the current layer. We start at -*Pad*, advance by *Stride* as the current layer advances, and end at twice the *HalfWidth.*

```
sum = 0.0;          // Will sum the filter here
// Center of first filter is at HalfWidth-Pad; filter begins at -Pad.
rstart = strideV[ilayer] * iheight - padV[ilayer];
rstop = rstart + 2 * HalfWidV[ilayer];
cstart = strideH[ilayer] * iwidth - padH[ilayer];
cstop = cstart + 2 * HalfWidH[ilayer];
```

We are now ready to compute the weighted sum of the prior layer's activations in the rectangle. Recall that the filter sums across all slices in the prior layer. Astute readers, and even not-so-astute readers, will notice a small but significant inefficiency in how I program the logic for handling zero padding outside the edges of the prior layer. The row test can be done outside the column loop since its result will be the same for all columns! However, I deliberately did it this way here for clarity. It should be trivial for interested readers to fix this. It is also possible to limit the rectangle bounds in advance so that no test is necessary. But that complicates weight addressing a lot and likely would be no faster.

```
for (in_slice=0; in_slice<in_slices; in_slice++) {
  for (in_row=rstart; in_row<=rstop; in_row++) {
    for (in_col=cstart; in_col<=cstop; in_col++) {

      // This logic is a bit inefficient

      if (in_row >= 0 && in_row < in_rows && in_col >= 0 && in_col < in_cols)
        x = inptr[(in_slice*in_rows+in_row)*in_cols+in_col];
      else            // We are outside the visual field, in the zero padded area
        x = 0.0;

      sum += x * *wtptr++;          // Equation 1-1 on Page 3

    } // For in_col
  } // For in_row
} // For in_slice
```

```
        sum += *wtptr++;                 // Bias in Equation 1-1
        sum = exp (2.0 * sum);           // Hyperbolic tangent activation function
        sum = (sum - 1.0) / (sum + 1.0); // Equation 1-4 on Page 3
        outptr[k++] = sum;
        } // For iwidth
      } // For iheight
    } // For idepth
}
```

Activating a Convolutional Layer

The code for activating a convolutional layer is almost identical to that for activating a locally connected layer. This is because the only difference between the two is that in a convolutional layer, for a given slice, all neurons in the visual field share the same set of weights. In a more general locally connected layer, the neurons all have their own weight sets.

For this reason, it's a borderline waste of trees to reproduce the code here. Still, I think it's instructive to compare them. I suggest that you flip pages back and forth, comparing the two algorithms. I'll jump right in, stopping only to point out the salient differences.

First, let's again consider how the weights that connect the prior layer to this convolutional layer are ordered. This is identical to the locally connected ordering, except that the height and width are omitted because the weights are the same for every neuron in the visual field.

> Current layer depth
>   Prior layer depth
>   Prior layer height
>   Prior layer width
>   Bias

```
void Model::activity_conv_no_thr (int ilayer, double *input)
{
  int k, in_row, in_rows, in_col, in_cols, in_slice, in_slices, iheight, iwidth, idepth;
  int rstart, rstop, cstart, cstop;
  double sum, *wtptr, *inptr, *outptr, x;
```

```
if (ilayer == 0) {
  in_rows = IMAGE_rows;
  in_cols = IMAGE_cols;
  in_slices = IMAGE_bands;
  inptr = input;
  }

else {
  in_rows = height[ilayer-1];
  in_cols = width[ilayer-1];
  in_slices = depth[ilayer-1];
  inptr = activity[ilayer-1];
  }
```

Here's the first difference. In the locally connected version, we initialized wtptr to the current layer's weight vector here, and it was incremented throughout the following (idepth, iheight, iwidth) nested loops because every neuron in the current layer had its own set of weights. But in a convolutional layer, each slice has its own weight set, with all neurons in the visual field of that slice sharing the same weights.

```
outptr = activity[ilayer];

k = 0;         // Will index computed activations in outptr
for (idepth=0; idepth<depth[ilayer]; idepth++) {
  for (iheight=0; iheight<height[ilayer]; iheight++) {
    for (iwidth=0; iwidth<width[ilayer]; iwidth++) {

      // Compute activation of this layer's neuron at (idepth, iheight, iwidth)
      // The weights for this layer are the same for all neurons in the layer's visual field
      // but a different such set is used for each slice
```

Here's the other difference, again having to do with the weights. Because every neuron in the visual field of a slice shares the same weight set, we must reset the weight pointer for each row and column. Past this point, everything is the same.

```
      wtptr = layer_weights[ilayer] + idepth * n_prior_weights[ilayer];

      sum = 0.0;
      // Center of first filter is at HalfWidth-Pad; filter begins at -Pad.
      rstart = strideV[ilayer] * iheight - padV[ilayer];
```

```
      rstop = rstart + 2 * HalfWidV[ilayer];
      cstart = strideH[ilayer] * iwidth - padH[ilayer];
      cstop = cstart + 2 * HalfWidH[ilayer];

        for (in_slice=0; in_slice<in_slices; in_slice++) {
          for (in_row=rstart; in_row<=rstop; in_row++) {
            for (in_col=cstart; in_col<=cstop; in_col++) {

              // This logic is a bit inefficient; see the CUDA implementation for better
              if (in_row >= 0 && in_row < in_rows && in_col >= 0 && in_col < in_cols)
                x = inptr[(in_slice*in_rows+in_row)*in_cols+in_col];
              else
                x = 0.0;
              sum += x * *wtptr++;

              } // For in_col
            } // For in_row
          } // For in_slice

        sum += *wtptr++;                     // Bias in Equation 1-1
        sum = exp (2.0 * sum);               // Hyperbolic tangent activation function
        sum = (sum - 1.0) / (sum + 1.0);   // Equation 1-4 on Page 3
        outptr[k++] = sum;
        } // For iwidth
      } // For iheight
    } // For idepth
  }
```

Activating a Pooling Layer

A pooling layer has no trainable weights. Like locally connected and convolutional layers, it moves a window across the prior layer to compute the activations of its neurons. However, the function that maps activations in a prior-layer window to a neuron in the current layer is fixed in advance. The sole purpose of a pooling layer is to efficiently reduce the visual-field resolution while preserving as much information as possible.

This text presents the two most popular types of pooling layers: average and max. Others exist but are not yet in widespread use. As in the previous layer types, the input parameter is the model's input image, used only in the rare circumstance that the first hidden layer is a pooling layer. The code starts just like that for earlier layers, gathering essential parameters and identifying the source of prior-layer activations.

```
void Model::activity_pool_no_thr (int ilayer, double *input)
{
  int k, in_row, in_rows, in_col, in_cols, in_slices, iheight, iwidth, idepth;
  int pwH, pwV, strH, strV, rstart, rstop, cstart, cstop;
  double value, *inptr, *outptr, x;

  pwH = PoolWidH[ilayer];           // Pooling width
  pwV = PoolWidV[ilayer];
  strH = strideH[ilayer];           // Stride
  strV = strideV[ilayer];

  if (ilayer == 0) {                // This is the first hidden layer (rare for pooling)
    in_rows = IMAGE_rows;
    in_cols = IMAGE_cols;
    in_slices = IMAGE_bands;
    inptr = input;
    }
  else {
    in_rows = height[ilayer-1];
    in_cols = width[ilayer-1];
    in_slices = depth[ilayer-1];
    inptr = activity[ilayer-1];
    }

  outptr = activity[ilayer];        // Computed activations will go here

  k = 0;                            // Will index computed activations in outptr
  for (idepth=0; idepth<depth[ilayer]; idepth++) { // Each prior-layer slice has slice here

    for (iheight=0; iheight<height[ilayer]; iheight++) {
      for (iwidth=0; iwidth<width[ilayer]; iwidth++) {
```

```
// Compute activation of this layer's neuron at (idepth, iheight, iwidth)
// Pooling layers never have padding, so we do not have to worry about
// logic for determining if we are outside the prior layer's visual field

rstart = strV * iheight;
rstop = rstart + pwV - 1;
cstart = strH * iwidth;
cstop = cstart + pwH - 1;
```

One type of pooling we can do is to simply take the average of the activations in the window. This was the original pooling, but it has fallen from favor recently.

```
if (layer_type[ilayer] == TYPE_POOLAVG) {
  value = 0.0;
  for (in_row=rstart; in_row<=rstop; in_row++) {
    for (in_col=cstart; in_col<=cstop; in_col++)
      value += inptr[(idepth*in_rows+in_row)*in_cols+in_col];
    } // For in_row
  value /= pwV * pwH;
  }
```

The other type of pooling presented here is currently the most popular. We examine all prior-layer activations in the window and choose whichever is the largest. We also save in poolmax_id the position in the window of this maximum activation. This will prove handy later when we backpropagate delta.

```
else if (layer_type[ilayer] == TYPE_POOLMAX) {
  value = -1.e60;
  for (in_row=rstart; in_row<=rstop; in_row++) {
    for (in_col=cstart; in_col<=cstop; in_col++) {
      x = inptr[(idepth*in_rows+in_row)*in_cols+in_col];
      if (x > value) {
        value = x;
        poolmax_id[ilayer][k] = in_row * in_cols + in_col; // Save id of max
        }
      } // For in_col
    } // For in_row
  }
```

```
            outptr[k++] = value;          // Save this activation

        } // For iwidth
      } // For iheight
    } // For idepth
}
```

Evaluating the Criterion

As part of a training procedure, we regularly want to pass through the entire training set in order to evaluate the performance criterion for a trial set of model parameters. We will use Equation 1-11. Also, many developers want to impose a small weight penalty to discourage the training algorithm from producing "optimal" weights that are overly large. This is primarily because larger weights tend to create overfitting. Advanced training algorithms may want to evaluate over only part of the training set, which is why we have istart and istop parameters. Here is the code for computing the criterion:

```
double Model::trial_error_no_thr (int istart, int istop)
{
  int i, icase, imax, ilayer, ineuron, ivar, n_prior;
  double err, tot_err, *dptr, tmax, *wptr, wt, wpen;

  tot_err = 0.0; // Total error will be cumulated here

  for (icase=istart; icase<istop; icase++) {   // Do all cases requested by caller

    dptr = database + icase * n_db_cols;    // Point to this case
    trial_no_thr (dptr);
    err = 0.0;

    tmax = -1.e30;
    imax = 0;                              // Not needed; be clean
    for (i=0; i<n_classes; i++) {          // The true class is that having max target
                // This is more general than using a single integer class id,
                // as it allows for probability-based class membership
      pred[icase*n_classes+i] = output[i];    // Save for other routines
```

```
   if (dptr[n_pred+i] > tmax) {
      imax = i;
      tmax = dptr[n_pred+i];
      }
   }
err = -log (output[imax] + 1.e-30);        // Equation 1-11 on Page 17
tot_err += err;
} // for all cases
```

There are several things to note in the code just shown.

- We save the outputs for every case in pred. This is optional, but some specialized performance criteria routines (such as for computing a confusion matrix) may call this routine for the sole purpose of generating all predictions. If you don't need them saved, don't bother.

- For each case, we check all targets and find the one having largest value. This is the "true" class. All this checking, repeated every time this routine is called, is inefficient (although usually tiny compared to the time taken by the call to trial_no_thr()). I did it this way here to show exactly what's going on and also to allow use of this routine in advanced situations in which true class probabilities may evolve. Most users would be best off precomputing the class membership, which in fact is what I do in the CUDA implementation presented later.

The last step is to implement the optional weight penalty. This is straightforward, but I'll list it here just to reinforce the architecture of the model. The most important thing to note is that we do not include the bias in the weight penalty because forcing the bias to be small might prevent properly centering activations near zero. Some developers might want to include the bias.

```
wpen = TrainParams.wpen / n_all_weights;        // Normalize to per-weight
penalty = 0.0;
for (ilayer=0; ilayer<=n_layers; ilayer++) {        // Do all hidden layers, plus output
   wptr = layer_weights[ilayer];
   n_prior = n_prior_weights[ilayer];        // This is per neuron
```

```
if (ilayer == n_layers) {                                // Output layer
  for (ineuron=0; ineuron<n_classes; ineuron++) {
    for (ivar=0; ivar<n_prior-1; ivar++) {              // Do not include bias in penalty
      wt = wptr[ineuron*n_prior+ivar];
      penalty += wt * wt;                               // Penalty is sum of squares
      }
    }
  }

else if (layer_type[ilayer] == TYPE_FC) {              // Fully connected layer
  for (ineuron=0; ineuron<nhid[ilayer]; ineuron++) {
    for (ivar=0; ivar<n_prior-1; ivar++) {   // Do not include bias in penalty
      wt = wptr[ineuron*n_prior+ivar];
      penalty += wt * wt;
      }
    }
  }

else if (layer_type[ilayer] == TYPE_LOCAL) {          // Locally connected layer
  for (ineuron=0; ineuron<nhid[ilayer]; ineuron++) {
    for (ivar=0; ivar<n_prior-1; ivar++) { // Do not include bias in penalty
      wt = wptr[ineuron*n_prior+ivar];
      penalty += wt * wt;
      }
    }
  }

else if (layer_type[ilayer] == TYPE_CONV) {
  // For CONV layers, each depth has its own weight set,
  // but weights across visual field are identical
  for (ineuron=0; ineuron<depth[ilayer]; ineuron++) {
    for (ivar=0; ivar<n_prior-1; ivar++) {    // Do not include bias in penalty
      wt = wptr[ineuron*n_prior+ivar];
      penalty += wt * wt;
      }
    }
  }
}
```

```
  penalty *= wpen;
  return tot_err / ((istop - istart) * n_classes) + penalty;
}
```

Note that we divide the total log likelihood criterion by the number of cases and classes. This is not strictly necessary, but such normalization is nice, both for printing the criterion for users as well as putting it on par with any weight penalty.

Evaluating the Gradient

Now would be a good time to flip back to page 21 and review the general description of gradient computation. We will refer to the numbered steps during this presentation of the code.

```
double Model::grad_no_thr (int istart, int istop)
{
  int i, j, icase, ilayer, nprev, imax, n_prior, ineuron, ivar;
  double *dptr, error, *prevact, *gradptr, delta, *nextcoefs, tmax, *wptr, *gptr, wt, wpen;

  for (i=0; i<n_all_weights; i++)      // Zero gradient for summing
    gradient[i] = 0.0;                 // All layers are strung together here

  error = 0.0;                         // Will cumulate total error here for return to user
  for (icase=istart; icase<istop; icase++) {
    dptr = database + icase * n_db_cols;    // Point to this case

/*
  Cumulate error criterion
*/

    trial_no_thr (dptr);               // Step 2: Compute all activations

    tmax = -1.e30;
    imax = 0;                          // Not needed
    for (i=0; i<n_classes; i++) { // Find the true class as that having max target
                   // This is more general than using a single integer class id,
                   // as it allows for probability-based class membership
```

```
        if (dptr[n_pred+i] > tmax) {
          imax = i;
          tmax = dptr[n_pred+i];
          }

        // Delta is the (negative) deriv of cross entropy wrt input (logit) i
        // We flip the sign because we are minimizing
        // This is Step 3, compute delta and put it in this_delta
        this_delta[i] = dptr[n_pred+i] - output[i];        // Equation 1-12 on Page 18
        }
      error -= log (output[imax] + 1.e-30);                // Equation 1-11 on Page 17
/*
  Cumulate output gradient: Step 4
*/

      if (n_layers == 0) {                          // No hidden layer
        nprev = n_pred;                             // Number of inputs to the output layer
        prevact = dptr;                             // Point to this sample
        }
      else {
        nprev = nhid[n_layers-1];                   // The last hidden layer
        prevact = activity[n_layers-1];             // Point to layer feeding the output layer
        }
      gradptr = layer_gradient[n_layers];           // Point to output gradient
      for (i=0; i<n_classes; i++) {                 // For all output neurons
        delta = this_delta[i];                      // Neg deriv of criterion wrt logit
        for (j=0; j<nprev; j++)
          *gradptr++ += delta * prevact[j];         // Equation 1-14 on Page 19
        *gradptr++ += delta;                        // Bias activation is always 1
        }
/*
  Cumulate hidden gradients.
  Each of these calls also backprops delta from this_delta to prior_delta.
  This is why we also have a call to grad_no_thr_POOL, even though
  a pooled layer has no weights and hence no gradient.
  That call handles backpropping delta just like the other calls.
*/
```

The following ilayer loop marks steps 5 and 6, which are more like bookkeeping steps than actual computation. The following calls to grad_no_thr_? implement steps 7 and 9, and step 8 follows as the last item in the loop. The CUDA implementation follows the steps exactly, while the slight reordering here improves efficiency.

```
for (ilayer=n_layers-1; ilayer>=0; ilayer--) { // For each hidden layer, backwards

    if (layer_type[ilayer] == TYPE_FC)
        grad_no_thr_FC (icase, ilayer);            // Step 7 and 9

    else if (layer_type[ilayer] == TYPE_LOCAL)
        grad_no_thr_LOCAL (icase, ilayer);         // Step 7 and 9

    else if (layer_type[ilayer] == TYPE_CONV)
        grad_no_thr_CONV (icase, ilayer);          // Step 7 and 9

    else if (layer_type[ilayer] == TYPE_POOLAVG ||
             layer_type[ilayer] == TYPE_POOLMAX)
        grad_no_thr_POOL (ilayer); // POOL has no weights, but this backprops delta

    for (i=0; i<nhid[ilayer]; i++)      // These will be delta for the next layer back
        this_delta[i] = prior_delta[i];  // Step 8

    } // For all layers, working backwards

  } // for all cases
 for (i=0; i<n_all_weights; i++)
   gradient[i] /= (istop - istart) * n_classes;
```

At the end of the code just shown, we divide the gradient sum by the number of cases and the number of classes. This is because we do the same for the performance criterion as a form of optional but nice normalization.

The last step is to compute the weight penalty. This was already discussed in the prior section, but here we have one additional task. Because the penalty is the sum of the square of each weight, the derivative is twice the value of the weight. Subtract that from the gradient.

```
wpen = TrainParams.wpen / n_all_weights;
penalty = 0.0;
for (ilayer=0; ilayer<=n_layers; ilayer++) { // Do all hidden layers, plus final
```

```
wptr = layer_weights[ilayer];
gptr = layer_gradient[ilayer];
n_prior = n_prior_weights[ilayer];
if (ilayer == n_layers) {                              // Output layer
  for (ineuron=0; ineuron<n_classes; ineuron++) {
    for (ivar=0; ivar<n_prior-1; ivar++) {  // Do not include bias in penalty
      wt = wptr[ineuron*n_prior+ivar];
      penalty += wt * wt;
      gptr[ineuron*n_prior+ivar] -= 2.0 * wpen * wt;
      }
    }
  }
else if (layer_type[ilayer] == TYPE_FC) {          // Fully connected layer
  for (ineuron=0; ineuron<nhid[ilayer]; ineuron++) {
    for (ivar=0; ivar<n_prior-1; ivar++) {   // Do not include bias in penalty
      wt = wptr[ineuron*n_prior+ivar];
      penalty += wt * wt;
      gptr[ineuron*n_prior+ivar] -= 2.0 * wpen * wt;
      }
    }
  }
else if (layer_type[ilayer] == TYPE_LOCAL) {       // Locally connected layer
  for (ineuron=0; ineuron<nhid[ilayer]; ineuron++) {
    for (ivar=0; ivar<n_prior-1; ivar++) {    // Do not include bias in penalty
      wt = wptr[ineuron*n_prior+ivar];
      penalty += wt * wt;
      gptr[ineuron*n_prior+ivar] -= 2.0 * wpen * wt;
      }
    }
  }
else if (layer_type[ilayer] == TYPE_CONV) {        // Convolutional layer
  // For CONV layers, weights across visual field are identical for each slice
  for (ineuron=0; ineuron<depth[ilayer]; ineuron++) {
    for (ivar=0; ivar<n_prior-1; ivar++) { // Do not include bias in penalty
      wt = wptr[ineuron*n_prior+ivar];
```

```
        penalty += wt * wt;
        gptr[ineuron*n_prior+ivar] -= 2.0 * wpen * wt;
        }
      }
    }
  }
  penalty *= wpen;
  return error / ((istop - istart) * n_classes) + penalty; // Negative log likelihood
}
```

Gradient for a Fully Connected Layer

A fully connected layer has the easiest gradient calculation algorithm because one does not need to worry about moving a window around the prior layer. Every neuron in the prior layer connects to every neuron in the current layer.

In the following code, note that database and n_db_cols are global.

```
void Model::grad_no_thr_FC (int icase, int ilayer)
{
  int i, j, nthis, nnext;
  double *gradptr, delta, *prevact, *nextcoefs;

  nthis = nhid[ilayer];        // N of neurons in this hidden layer (height * width * depth)
  if (ilayer == n_layers-1)  // Next layer is output layer?
    nnext = n_classes;         // Number of neurons in next layer
  else                         // Next layer is another hidden layer
    nnext = nhid[ilayer+1];

  if (ilayer == 0)             // First hidden layer?
    prevact = database + icase * n_db_cols; // Point to this sample
  else                         // There is at least one more hidden layer prior to this one
    prevact = activity[ilayer-1];

  gradptr = layer_gradient[ilayer];       // Point to grad for this layer; will put results here
  nextcoefs = layer_weights[ilayer+1]; // Weights for the next layer are here
```

All of these gradient routines (but not the CUDA versions) implement steps 7 (backpropping delta) and then step 9 (gradient computation), letting the caller do step 8 (copy prior_delta to this_delta) later. Each of the nthis hidden neurons in this layer is processed individually. Within this loop, the first step is to see whether the next layer is a fully connected layer. Recall that the output layer is always fully connected. If fully connected, then the summation in Equation 1-19 is trivial. We just sum delta over the nnext neurons in the next layer.

```
for (i=0; i<nthis; i++) {      // For each neuron in this layer

  if (ilayer+1 == n_layers || layer_type[ilayer+1] == TYPE_FC) { // Simple; just sum
    delta = 0.0;
    for (j=0; j<nnext; j++)
      delta += this_delta[j] * nextcoefs[j*(nthis+1)+i]; // The +1 is for the bias term
    }
```

But if the next layer is anything other than fully connected, backpropagating delta is a lot more complicated than just summing all connections; we have a moving window to deal with. So, we call a subroutine to do it. We have two such routines, one for locally connected and convolutional layers (nonpooled) and one for pooled layers. These two subroutines compute all deltas simultaneously. Thus, we call them for only the first pass through the neuron loop, i=0. For subsequent neurons, we just fetch delta from the array that was computed for the first neuron.

```
  else if (i == 0) {  // Will compute all deltas at once
    if (layer_type[ilayer+1] == TYPE_LOCAL || layer_type[ilayer+1] == TYPE_CONV)
      compute_nonpooled_delta (ilayer);
    else if (layer_type[ilayer+1] == TYPE_POOLAVG ||
            layer_type[ilayer+1] == TYPE_POOLMAX)
      compute_pooled_delta (ilayer);
    delta = prior_delta[i];
    }
  else                          // We're past the first neuron
    delta = prior_delta[i];     // Delta is already computed (just above) and saved
```

We still have to multiply the sum by the derivative of the activation function (Equation 1-15) to complete Equation 1-19. We do that and save the result in prior_delta.

```
  delta *= 1.0 - activity[ilayer][i] * activity[ilayer][i];   // Eq (1.15) finishes Eq (1.19)
  prior_delta[i] = delta;                                      // Save it for the next layer back
```

Finally, compute the gradient using Equation 1-20.

```
for (j=0; j<n_prior_weights[ilayer]-1; j++)      // Don't include bias yet
   *gradptr++ += delta * prevact[j];             // Equation 1-20 on Page 21
 *gradptr++ += delta;                            // Bias activation is always 1
 } // For all neurons in this hidden layer
}
```

Gradient for a Locally Connected Layer

In terms of what we are actually doing, computation of the gradient of a locally connected layer is the same as for a fully connected layer. The hitch is that for a locally connected layer, most of the connections from the prior layer to the current layer are zero; only the weights in each window are nonzero. It is vital that we have an efficient way to process only the nonzero weights.

Much of this code is similar to that in the prior section, so explanations of those parts will be omitted. The only early difference is that we now need the dimensions of the prior layer.

```
void Model::grad_no_thr_LOCAL (int icase, int ilayer)
{
  int j, k, nthis, nnext, idepth, iheight, iwidth;
  int in_row, in_col, in_slice, in_rows, in_cols, in_slices;
  int rstart, rstop, cstart, cstop;
  double *gradptr, delta, *prevact, *nextcoefs, x;

  nthis = nhid[ilayer];        // N of neurons in this hidden layer (height * width * depth)
  if (ilayer == n_layers-1)   // Next layer is output layer?
    nnext = n_classes;
  else
    nnext = nhid[ilayer+1];

  if (ilayer == 0) {
    prevact = database + icase * n_db_cols;        // Point to this case
    in_rows = IMAGE_rows;                          // These, too, are global
    in_cols = IMAGE_cols;
    in_slices = IMAGE_bands;
    }
```

```
else {
  prevact = activity[ilayer-1];
  in_rows = height[ilayer-1];
  in_cols = width[ilayer-1];
  in_slices = depth[ilayer-1];
  }
gradptr = layer_gradient[ilayer];              // Point to gradient for this layer
nextcoefs = layer_weights[ilayer+1];           // Weights for next layer
```

For the fully connected layer discussed in the prior section, we looped over all neurons in the current layer. We do the same here, except that now we must break it into each dimension separately.

```
k = 0; // This will index the nhid[ilayer] neurons in this layer
for (idepth=0; idepth<depth[ilayer]; idepth++) {
  for (iheight=0; iheight<height[ilayer]; iheight++) {
    for (iwidth=0; iwidth<width[ilayer]; iwidth++) {

      //-----------------------------------------------------------------------------
      // We are now inside the three nested loops that cover all nhid[ilayer]
      // neurons in this layer. Compute delta for this neuron by summing
      // across all connections to the next layer.
      //-----------------------------------------------------------------------------
```

Exactly as in the fully connected case, we do simple summation across all neurons in the next layer. But for locally connected and convolutional next layers, we must call the specialized subroutine that computes all deltas.

```
if (ilayer+1 == n_layers || layer_type[ilayer+1] == TYPE_FC) { // Simple case
  delta = 0.0;
  for (j=0; j<nnext; j++)
    delta += this_delta[j] * nextcoefs[j*(nthis+1)+k];
  }
```

```
else if (idepth == 0 && iheight == 0 && iwidth == 0) { // Will compute all deltas
  if (layer_type[ilayer+1] == TYPE_LOCAL ||
     layer_type[ilayer+1] == TYPE_CONV)
    compute_nonpooled_delta (ilayer);
  else if (layer_type[ilayer+1] == TYPE_POOLAVG ||
        layer_type[ilayer+1] == TYPE_POOLMAX)
    compute_pooled_delta (ilayer);
  delta = prior_delta[k];
  }

else
  delta = prior_delta[k]; // It's already computed (just above) and saved
  // At this point, delta for this layer's hidden neuron k at (idepth, iheight, iwidth)
  // is the derivative of the criterion wrt the output of this hidden neuron.
  // To make it be wrt the input to this neuron, multiply by the derivative
  // of the activation function.
  // Note that this multiplication takes place only once for each neuron k.

  delta *= 1.0 - activity[ilayer][k] * activity[ilayer][k]; // Eq (1.15) finishes Eq (1.19)
  prior_delta[k] = delta; // Save it for the next layer back
                // Delta is now the derivative of the crit wrt net input to neuron k
```

To get the gradient, we use Equation 1-20. The method for computing the location of the current neuron's rectangle in the prior layer is exactly as described in the section on computing activation of this neuron, page 31, so it won't be repeated here. While you're in that section, please review the order of neurons in a layer and the layout of the gradient vector.

Also, this code uses the same inefficient but clear logic of needlessly checking the row bounds for every column. The extra time is a tiny fraction of the total time, but many readers will want to fix it. Note that the CUDA code presented later does it efficiently.

```
//-----------------------------------------------------------------------------
// To get the derivative of the criterion with respect to the
// n_prior_weights coming into this neuron, multiply delta
// by the corresponding input to the weight.
//-----------------------------------------------------------------------------
```

```
      // Center of first filter is at HalfWidth-Pad; filter begins at -Pad.
      rstart = strideV[ilayer] * iheight - padV[ilayer];
      rstop = rstart + 2 * HalfWidV[ilayer];
      cstart = strideH[ilayer] * iwidth - padH[ilayer];
      cstop = cstart + 2 * HalfWidH[ilayer];
   for (in_slice=0; in_slice<in_slices; in_slice++) {
     for (in_row=rstart; in_row<=rstop; in_row++) {
       for (in_col=cstart; in_col<=cstop; in_col++) {

         // This logic is a bit inefficient
         if (in_row >= 0 && in_row < in_rows && in_col >= 0 && in_col < in_cols)
           x = prevact[(in_slice*in_rows+in_row)*in_cols+in_col];
         else
           x = 0.0;
         *gradptr++ += delta * x;

       } // For every column in the prior layer
     } // For every row in the prior layer
   } // For every slice in the prior layer

   *gradptr++ += delta; // Bias activation is always 1
   ++k;

   } // For width dimension in this hidden layer
  } // For height dimension in this hidden layer
 } // For depth dimension in this hidden layer
}
```

Gradient for a Convolutional Layer

The code for computing the gradient for a convolutional layer is almost exactly the same as the code for a locally connected layer. The only difference is that a locally connected layer has a separate weight set for every hidden neuron, so gradptr is set at the start of processing and incremented throughout. However, a convolutional layer uses the same weight set for all neurons in the visual field of a given slice. Thus, we reset gradptr according to the current slice every time we begin processing a new neuron in the visual field. Here is this change, shown in context. All other code is the same for both layer types and hence omitted here.

```
delta *= 1.0 - activity[ilayer][k] * activity[ilayer][k];   // Eq (1.15) finishes Eq (1.19)
prior_delta[k] = delta;    // Save it for the next layer back
                    // Delta is the derivative of the crit wrt net input to neuron k

//-------------------------------------------------------------------------------
// To get the derivative of the criterion with respect to the
// n_prior_weights coming into this neuron, multiply delta
// by the corresponding input to the weight.
//-------------------------------------------------------------------------------

// Weights for this layer are the same for all neurons in the visible field
// But a different set is used for each slice in this layer
// The line below is the only difference between this code and that
// for a locally connected layer.

gradptr = layer_gradient[ilayer] + idepth * n_prior_weights[ilayer];

// Center of first filter is at HalfWidth-Pad; filter begins at -Pad.
rstart = strideV[ilayer] * iheight - padV[ilayer];
rstop = rstart + 2 * HalfWidV[ilayer];
cstart = strideH[ilayer] * iwidth - padH[ilayer];
cstop = cstart + 2 * HalfWidH[ilayer];
```

Gradient for a Pooled Layer (Not!)

In the section on general gradient computation (page 42) you may have noticed a call to subroutine grad_no_thr_POOL(). On the surface, this seems rather silly, as a pooled layer is a fixed function, and hence it has neither trainable weights nor a gradient. It is, nevertheless, a functional layer and hence plays a role in both forward activation and delta backpropagation. The set of specialized routines presented in the past few sections all perform two duties: they backpropagate delta, and they compute the gradient. To preserve the structure, I included grad_no_thr_POOL(), which has the single duty of handling backpropagation.

There is no point in showing the code for this routine here. It is, in essence, the first part of the two prior routines that handle locally connected and convolutional layers. This code just organizes the backpropagation as shows earlier and stops before computing the nonexistent gradient. Naturally, this code can be found in the source files able to be downloaded from my web site.

Backpropagating Delta from a Nonpooled Layer

The specialized gradient routines shown in the previous few sections directly backpropagate delta in the simple case that the next layer is fully connected. However, other layer types call a specialized backpropagation routine. The one that handles locally connected and convolutional layers is presented in this section.

A potentially confusing reversal of loop nesting happens in this algorithm. Look back at Equation 1-19, and review the discussion of backpropagation that precedes this equation if necessary. For a given neuron in the current layer, the summation is over connections to the *next* layer. However, as should be clear by now from the sections on activation and gradient computation, connections are defined between a neuron and its associated rectangle in the *prior* layer. For a given neuron, it's easy to define the neurons in the *prior* layer to which it connects. On the other hand, it can be quite difficult to define, and inefficient to compute, the connections from a given layer to the *next* layer. Unfortunately, this is precisely what a superficial implementation of Equation 1-19 requires.

To circumvent this problem, we reverse the order of summation in this equation, which implies that we must compute all deltas simultaneously. In other words, we zero all deltas before beginning. Then we have an outer set of loops over neurons in the *next* layer, and an inner set of loops over neurons in the *current* layer. As each connection is processed, update the associated delta. Thus, the summation of Equation 1-19 is split into many parts, cumulated in widely separated passes. Ideally, this will become clearer after studying the code.

```
void Model::compute_nonpooled_delta (int ilayer)
{
  int i, next_row, next_col, next_slice, next_rows, next_cols, next_slices;
  int this_slices, this_rows, this_cols, idepth, iheight, iwidth;
  int hwH, nH, hwV, nV, pdH, pdV, rstart, rstop, cstart, cstop, strH, strV, k_this, k_next;
  double *wtptr;

  for (i=0; i<nhid[ilayer]; i++)        // Zero all deltas before beginning
    prior_delta[i] = 0.0;

  hwH = HalfWidH[ilayer+1];        // Filter half-width in next layer
  nH = 2 * hwH + 1;               // And its number of columns
  hwV = HalfWidV[ilayer+1];        // Ditto for rows
```

```
nV = 2 * hwV + 1;
strH = strideH[ilayer+1];
strV = strideV[ilayer+1];
pdH = padH[ilayer+1];
pdV = padV[ilayer+1];

this_rows = height[ilayer];
this_cols = width[ilayer];
this_slices = depth[ilayer];

next_rows = height[ilayer+1];
next_cols = width[ilayer+1];
next_slices = depth[ilayer+1];
```

```
/*
   Loop through every possible connection from a neuron in ilayer
   to a neuron in the next layer. This is a loop reversal from Equation 1-19.
   In that equation, we pick a neuron in the current layer and loop over
   connections to the next layer. But here we pick a neuron in the next layer
   and loop over the current layer (which the next layer's prior layer).
*/
```

```
k_next= 0; // Will index every neuron in the next layer
for (next_slice=0; next_slice<next_slices; next_slice++) {
  for (next_row=0; next_row<next_rows; next_row++) {
    for (next_col=0; next_col<next_cols; next_col++) {
```

We now point to the weights connecting this "next" layer to its "prior" layer, which we might call the *current* layer. A convolutional layer has the same weight set for all neurons in the visual field of a given slice, while a locally connected layer has different weights for each neuron.

```
      if (layer_type[ilayer+1] == TYPE_CONV)
        wtptr = layer_weights[ilayer+1] + next_slice * n_prior_weights[ilayer+1];
      else if (layer_type[ilayer+1] == TYPE_LOCAL)
        wtptr = layer_weights[ilayer+1] + k_next * n_prior_weights[ilayer+1];
      else
        wtptr = NULL;   // Not needed. Shuts up picky compilers.
```

Here we have the old, familiar bounding rectangle. We also have the same inefficient but clear excessive row checking, which picky readers will revise. Again, the CUDA implementation does it better.

```
// Center of first filter is at HalfWidth-Pad; filter begins at -Pad.
rstart = strV * next_row - pdV;
rstop = rstart + 2 * hwV;
cstart = strH * next_col - pdH;
cstop = cstart + 2 * hwH;

for (idepth=0; idepth<this_slices; idepth++) {
  for (iheight=rstart; iheight<=rstop; iheight++) {
    for (iwidth=cstart; iwidth<=cstop; iwidth++) {

      if (iheight >= 0 && iheight < this_rows &&
        iwidth >= 0 && iwidth < this_cols) {
        k_this = (idepth * this_rows + iheight) * this_cols + iwidth;
        prior_delta[k_this] += this_delta[k_next] * *wtptr++;
        }
      else
        ++wtptr;

      } // For iwidth
    } // For iheight
  } // For idepth

  ++k_next;

  } // For next_col
  } // For next_row
  } // For next_slice
}
```

Ideally, the concept of loop reversal is clear now. Instead of picking one neuron at a time and summing Equation 1-19 over the next layer, we pick from the next layer one term of a sum at a time and compute (k_this) the particular sum to which this term belongs. This method is much more efficient than naive computation of each sum, which requires complex logic.

Backpropagating Delta from a Pooled Layer

When we backpropagate delta from a pooled layer, we do the same loop reversal that we did for a nonpooled layer. In fact, the algorithm here is similar to the algorithm presented in the prior section. It's somewhat easier, though, because pooled layers are never padded (at least not by me), which means we do not have to check for the rectangle extending over the edge of the prior layer's visual field. We begin by zeroing all deltas and then fetching some constants that will be referenced often later.

```
void Model::compute_pooled_delta (int ilayer)
{
  int i, pwH, pwV, next_row, next_col, next_slice, next_rows, next_cols, next_slices;
  int this_slices, this_rows, this_cols, iheight, iwidth;
  int rstart, rstop, cstart, cstop, strH, strV, k_this, k_next;
  double wt;

  for (i=0; i<nhid[ilayer]; i++)
    prior_delta[i] = 0.0;

  pwH = PoolWidH[ilayer+1]; // Pooling filter width in next layer
  pwV = PoolWidV[ilayer+1];
  strH = strideH[ilayer+1];
  strV = strideV[ilayer+1];

  this_rows = height[ilayer];
  this_cols = width[ilayer];
  this_slices = depth[ilayer];

  next_rows = height[ilayer+1];
  next_cols = width[ilayer+1];
  next_slices = depth[ilayer+1];
```

As we did in the prior section, the outer loop here is what would be the inner loop in Equation 1-19. The counter k_next indexes neurons (and hence this_delta) in the next layer.

```
  k_next= 0; // Will index every neuron in the next layer
  for (next_slice=0; next_slice<next_slices; next_slice++) {
    for (next_row=0; next_row<next_rows; next_row++) {
      for (next_col=0; next_col<next_cols; next_col++) {
```

If this pooled layer is the "average" type, we find the bounding rectangle and process every connection in it. Note that the bounding rectangle here is considerably simpler than the bounding rectangle for locally connected and convolutional layers. This is because there is no padding. We also compute wt, the effective weight that went into computing the activation. Recall that when we computed the pooled average, we just divided the sum by the number of neurons going into the sum.

Also note that wt is a constant. I put the multiplication where it is most clear. However, it is somewhat inefficient to do all that multiplication deep inside a bunch of nested loops. Many readers will want to remove that multiplication from where it is and then just do it to each prior_delta at the end, outside all loops.

```
if (layer_type[ilayer+1] == TYPE_POOLAVG){
  wt = 1.0 / (pwH * pwV);
  rstart = strV * next_row;
  rstop = rstart + pwV - 1;
  cstart = strH * next_col;
  cstop = cstart + pwH - 1;

  for (iheight=rstart; iheight<=rstop; iheight++) {
    for (iwidth=cstart; iwidth<=cstop; iwidth++) {
      k_this = (next_slice * this_rows + iheight) * this_cols + iwidth;
      prior_delta[k_this] += this_delta[k_next] * wt;
      } // For iwidth
    } // For iheight
  } // If POOLAVG
```

Now we look at max pooling. In this type of pooling, we check each prior-layer neuron in the window and choose the one having maximum activation. This was discussed in the section that begins on page 36. The activation function saved the index of this winning neuron. We now decode this saved identity, getting the row as iheight and the column as iwidth. The "weight" of this connection is 1.0 because it is an exact copy. The weight of all other neurons in the rectangle is zero.

```
else if (layer_type[ilayer+1] == TYPE_POOLMAX) {
  iheight = poolmax_id[ilayer+1][k_next] / this_cols;
  iwidth = poolmax_id[ilayer+1][k_next] % this_cols;
  k_this = (next_slice * this_rows + iheight) * this_cols + iwidth;
  prior_delta[k_this] += this_delta[k_next]; // Weight is 1
  }
```

```
      ++k_next;

    } // For next_col
  } // For next_row
 } // For next_slice
}
```

Multithreading Gradient Computation

The source code that can be downloaded from my web site includes threaded versions of both criterion and gradient computation; these are in file MOD_THR.CPP. However, the criterion algorithm is just a subset of the gradient algorithm, so we will present only the gradient version here.

One thing that makes multithreaded computation a bit more difficult than single-thread code is that when a threaded routine is launched, you can pass only one parameter to the routine. So, you'd better make it a good one. The usual method is to define a data structure that contains everything the routine needs, put everything into that structure, and then pass a pointer to it as the sole legal argument.

Although it is possible to run class member functions in threaded mode, this is fraught with a wide assortment of gotchas. So I always prefer to take the old but safer route of making every function stand-alone, with all required parameters passed in a long parameter list. It's ugly, but you are a lot less likely to be stuck with a bizarre runtime error that can be horrendous to debug.

Here is the data structure that encapsulates everything that gradient computation needs. I made sure to give them names identical to Model class names as much as possible to reduce confusion.

```
typedef struct {
   int istart;              // Index of first case in batch
   int istop;               // And one past last case
   int n_all_weights;       // Includes bias and final layer weights
   double *gradient;        // 'n_all_weights' gradient; aligned with weights
   int n_layers;            // N of hidden layers; does not include input or output layer
   int *layer_type;         // Type of each layer
   double *output;          // Put the computed outputs here
   double **activity;       // Activity vector for each layer; used only when ilayer>0
```

```
int *HalfWidH;          // Horizontal half width looking back to prior layer
int *HalfWidV;          // And vertical
int *padH;              // Horizontal padding; must not exceed half width
int *padV;              // And vertical
int *strideH;           // Horizontal stride
int *strideV;           // And vertical
int *PoolWidH;          // Horizontal half width looking back to prior layer
int *PoolWidV;          // And vertical
double **layer_weights; // Pointers to each layer's weights in 'weight' vector
double **layer_gradient; // Pointers to each layer's gradient in 'gradient' vector
int *height;            // N of neurons vertically in a slice of this layer
int *width;             // Ditto horizontal
int *depth;             // Number of slices in this layer
int *nhid;              // Total number of neurons in this layer = H * W * D
double *this_delta;     // Scratch vector for gradient computation
double *prior_delta;    // Ditto
int **poolmax_id;       // Used only for POOLMAX layer; saves ID of max
int *n_prior_weights;   // N of inputs per neuron (including bias) to prior layer
double error;           // performance criterion is returned here
} GRAD_PARAMS;
```

After the members of this data structure have been filled in, a thread runs the routine shown next. Most of the interior of the parameter list is omitted for clarity. Note that this routine has a single parameter, dp, and it calls the real worker, batch_grad(). This latter routine is essentially identical to the grad_no_thr() routine presented on page 42. The only difference is that this routine cannot reference any model variables. Instead, everything must be passed to it in the long parameter list. (Well, it does reference several read-only globals, such as the database. See MOD_THR.CPP for details. It's straightforward, I promise.)

```
static unsigned int __stdcall batch_grad_wrapper (LPVOID dp)
{
((GRAD_PARAMS *) dp)->error = batch_grad (
  ((GRAD_PARAMS *) dp)->istart,
  ((GRAD_PARAMS *) dp)->istop,
  ((GRAD_PARAMS *) dp)->n_all_weights,
  ...
```

```
        ((GRAD_PARAMS *) dp)->poolmax_id,
        ((GRAD_PARAMS *) dp)->n_prior_weights);
     return 0;
}
```

This brings us to the nuts-and-bolts part of this multithreading presentation. Here is the Model member function that computes the gradient by running multiple threads simultaneously. The first step is to fill in the data structure as much as we can right now.

```
double Model::grad_thr (int jstart, int jstop)
{
   int i, nc, ret_val, ithread, n_threads, n_in_batch, n_done, istart, istop;
   int ilayer, ineuron, ivar, n_prior;
   double error, wpen, wt, *wptr, *gptr;
   GRAD_PARAMS params[MAX_THREADS];
   HANDLE threads[MAX_THREADS];

   nc = jstop - jstart;                    // Number of cases
   for (i=0; i<max_threads; i++) {    // max_threads may be up to MAX_THREADS
     params[i].n_all_weights = n_all_weights;
     params[i].gradient = thr_gradient[i]; // Each is allocated n_all_weights long
     params[i].n_layers = n_layers;
     params[i].layer_type = layer_type;
     params[i].output = thr_output + i * n_classes; // Allocated n_classes*max_threads
     params[i].activity = thr_activity[i]; // See Page 63 for allocation
     params[i].HalfWidH = HalfWidH;
     params[i].HalfWidV = HalfWidV;
     params[i].padH = padH;
     params[i].padV = padV;
     params[i].strideH = strideH;
     params[i].strideV = strideV;
     params[i].PoolWidH = PoolWidH;
     params[i].PoolWidV = PoolWidV;
     params[i].layer_weights = layer_weights;
     params[i].layer_gradient = thr_layer_gradient[i];     // See Page 63 for allocation
     params[i].height = height;
     params[i].width = width;
```

```
    params[i].depth = depth;
    params[i].nhid = nhid;
    params[i].this_delta=thr_this_delta+i*max_any_layer; //max_any_layer*max_threads
    params[i].prior_delta = thr_prior_delta + i * max_any_layer; // Ditto
    params[i].poolmax_id = thr_poolmax_id[i]; // See Page 63 for allocation
    params[i].n_prior_weights = n_prior_weights;
    }
```

Several of the parameters that go into the data structure are somewhat complicated because they are work areas that must not be shared among threads; each thread needs its own private copy so that they do not interfere with one another. These allocations are shown in the section that begins on page 63.

We will split up the training set into subsets that will be processed simultaneously by multiple threads. Launching a thread involves significant overhead, so we use an arbitrary rule (feel free to change it) to set the number of threads.

```
n_threads = nc / 100;           // This is the number of threads that we will launch
if (n_threads < 1)              // Division by 100 is arbitrary; change 100 at will
    n_threads = 1;
if (n_threads > max_threads)
    n_threads = max_threads;

istart = jstart;                // Batch start = training data start
n_done = 0;                     // Number of training cases done so far
```

This is the loop that launches all threads simultaneously. We use istart and istop to delineate the bounds of the subset being launched. The size of each launch (n_in_batch) is the number of training set cases left to do, divided by the number of threads left to process batches.

```
for (ithread=0; ithread<n_threads; ithread++) {
    n_in_batch = (nc - n_done) / (n_threads - ithread); // Cases left / batches left
    istop = istart + n_in_batch;                         // Stop just before this index

    // Set the pointers that vary with the batch

    params[ithread].istart = istart;    // The ithread batch will process this range of cases
    params[ithread].istop = istop;
```

```
// This is the Windows API call that launches the thread

threads[ithread] = (HANDLE) _beginthreadex (NULL, 0, batch_grad_wrapper,
                                            &params[ithread], 0, NULL);
```

It would be extremely unusual for the launch to fail, but a responsible programmer handles this possibility.

```
if (threads[ithread] == NULL) {
  // Post an error message here
  for (i=0; i<n_threads; i++) {
    if (threads[i] != NULL)
      CloseHandle (threads[i]);          // Clean up after yourself
    }
  return -1.e40;                          // Return an error flag to the caller
  }

n_done += n_in_batch;                     // Update number of cases running
istart = istop;                           // Advance to the next batch
} // For all threads / batches
```

The threads are running. Now we sit right here and wait until they are all finished. The time parameter, 1200000, is arbitrary but must be large enough to handle huge problems yet small enough that users don't give up and reboot. As in the launch, failure here is highly unlikely, but we must prepare for it.

```
ret_val = WaitForMultipleObjects (n_threads, threads, TRUE, 1200000);
if (ret_val == WAIT_TIMEOUT || ret_val == WAIT_FAILED ||
    ret_val < 0 || ret_val >= n_threads) {
  // Issue a general error message here
  if (ret_val == WAIT_TIMEOUT)
    // A 'problem too large' message may be appropriate here
  return -1.e40;          // Return an error flag to the caller
  }
```

All computation is done, and the results are in private areas of each thread. We will cumulate these results, so zero the sums here.

```
error = 0.0;                          // Cumulates performance criterion
for (i=0; i<n_all_weights; i++)       // Zero gradient for summing
   gradient[i] = 0.0;                 // All layers are strung together here
```

Here is where we add up the performance criterion and gradient for all threads and store them in the Model variables. As each thread's results are fetched, we close the thread. Finally, we normalize the gradient by dividing by the number of cases and classes. We will do this same division to the criterion at the end, when we return.

```
for (ithread=0; ithread<n_threads; ithread++) {
   error += params[ithread].error;
   for (i=0; i<n_all_weights; i++)
      gradient[i] += params[ithread].gradient[i];
   CloseHandle (threads[ithread]);
   }

for (i=0; i<n_all_weights; i++)
   gradient[i] /= nc * n_classes;
```

The last step is to handle the weight penalty. We won't bother showing this long stretch of code because we already saw it in conjunction with the nonthreaded criterion code. That section begins on page 39.

```
wpen = TrainParams.wpen / n_all_weights;
penalty = 0.0;
for (ilayer=0; ilayer<=n_layers; ilayer++) { // Do all hidden layers, plus final
   ...
   }

return error / (nc * n_classes) + penalty; // Negative log likelihood
}
```

Memory Allocation for Threading

As we saw a few pages back, the first thing done in the multithreaded version of gradient computation is to fill in the data structure that is passed to threads. Several of these entries are for work areas that must be private to each thread. Allocating some of them can be tricky, so this section will present code fragments that illustrate how to do this.

Here are the Model declarations of the four items discussed now:

```
double *thr_activity[MAX_THREADS][MAX_LAYERS];
int *thr_poolmax_id[MAX_THREADS][MAX_LAYERS];
double *thr_gradient[MAX_THREADS];
double *thr_layer_gradient[MAX_THREADS][MAX_LAYERS+1];
```

The two-dimensional arrays thr_activity and thr_poolmax_id are, for each value of the first dimension, exact analogs of the activity and poolmax_id member variables of the Model class. Every thread needs its own private copy, so this accounts for the first dimension. To implement this, we start by doing the master allocation and then split it up among the threads.

```
for (ilayer=0; ilayer<n_layers; ilayer++) {

  thr_activity[0][ilayer] = (double *) malloc (max_threads * nhid[ilayer] * sizeof(double));
  if (layer_type[ilayer] == TYPE_POOLMAX)
    thr_poolmax_id[0][ilayer] = (int *) malloc (max_threads * nhid[ilayer] * sizeof(int));

  for (i=1; i<max_threads; i++) {
    thr_activity[i][ilayer] = thr_activity[0][ilayer] + i * nhid[ilayer];
    if (layer_type[ilayer] == TYPE_POOLMAX)
      thr_poolmax_id[i][ilayer] = thr_poolmax_id[0][ilayer] + i * nhid[ilayer];
    }
  }
```

Because for each thread the gradient for all layers needs to be contiguous, we do things a little differently. We allocate the full gradient for each thread and then compute the position of each layer's gradient in this grand vector.

```
thr_gradient[0] = (double *) malloc (n_all_weights * max_threads * sizeof(double));

for (i=0; i<max_threads; i++) {
  k = 0;
  gptr = thr_gradient[0] + i * n_all_weights; // Gradient for this thread starts here
  thr_gradient[i] = gptr;

  for (ilayer=0;; ilayer++) {        // For each of the hidden layers, plus the final
    thr_layer_gradient[i][ilayer] = gptr + k;
    if (ilayer >= n_layers)          // Are we done?
      break;
```

```
   if (layer_type[ilayer] == TYPE_FC || layer_type[ilayer] == TYPE_LOCAL)
      k += nhid[ilayer] * n_prior_weights[ilayer]; // Add in weights for this layer
   else if (layer_type[ilayer] == TYPE_CONV)
      k += depth[ilayer] * n_prior_weights[ilayer]; // Convolution uses same per slice
   else if (layer_type[i] == TYPE_POOLAVG || layer_type[i] == TYPE_POOLMAX)
      k += 0;                        // Just for clarity; pooling has no trainable weights
   } // For ilayer
} // For i (thread)
```

CHAPTER 3

CUDA Code

The source code for the CUDA implementation of convolutional nets is in two files, both of which can be downloaded for free from my web site. MOD_CUDA.CPP provides the high-level organization. It calls subroutines to initialize, compute forward activation, backpropagate delta, and compute the gradient. MOD_CUDA.cu contains the CUDA device routines, as well as the low-level C++ host routines that are called from MOD_CUDA.CPP and that in turn launch computation kernels and provide communication between the host and the device.

Many excellent books on CUDA programming exist. It would be hopeless to try in this book to educate inexperienced readers in even the most basic aspects of CUDA programming. Volume 1 of my *Deep Belief Networks in C++ and CUDA C* series does contain an overview for the curious and uninitiated. However, this entire chapter will assume that you have at least modest competence in CUDA programming.

There are, however, several topics at what one might call the "advanced beginner" level that I will emphasize in the coding when appropriate, even though programmers at the intermediate level or beyond will be intimately familiar with these topics. These include the following:

- When doing large-scale accesses of global memory, it is crucial that, at a minimum, adjacent threads in a warp address adjacent memory addresses so that reads from the cache can be coalesced.

- In addition, it is even more profitable if memory accesses of the first thread in a warp are on an address that is divisible by 128 bytes. This allows full coalescing, matching warps with cache line blocks.

- Shared memory has much faster read access than global memory. Therefore, whenever possible one should do a single global memory read to shared memory and perform subsequent accesses from the shared memory.

© Timothy Masters 2018
T. Masters, *Deep Belief Nets in C++ and CUDA C: Volume 3*, https://doi.org/10.1007/978-1-4842-3721-2_3

- Especially with the most modern CUDA devices, it is almost always best to use a large number of relatively small blocks to give the scheduler maximum flexibility.

Weight Layout in the CUDA Implementation

On page 31 we saw how weights in memory on the host machine are organized for a locally connected layer, and on page 34 we saw the same for a convolutional layer. Please review those sections if needed. That layout facilitates the use of highly efficient dot product routines such as those described in Volume 1 of this series (though not shown in this volume). However, for reasons that will become clear later, that layout would be disastrous for a CUDA implementation.

On the device, the weights for a locally connected layer are organized as follows:

Input height
  Input width
    Input depth
Bias
      Layer height
        Layer width
          Layer depth
    Pad so nhid = layer height*width*depth is a multiple of 128 bytes

In a convolutional layer, which has identical weights for all neurons in the visual field of a given slice, or a fully connected layer, which has a 1×1 visual field, the organization is as follows:

Input height
  Input width
    Input depth
Bias
      Layer depth
      Pad so layer depth is a multiple of 128 bytes

If this is not clear, it should be made clearer on page 72 when the subject of copying host weights to the device is discussed. The most critical aspect of this layout is that weights along the depth dimension of the current layer change fastest, and they are padded to ensure full cache line coalescing.

Global Variables on the Device

Everything that any device routine may need is stored in globally accessible memory on the device, in constant memory whenever possible. Recall that constant memory occupies a special status that grants it very high-speed access. Moreover, if all threads in a warp access the same constant memory simultaneously, which is the usual case, speed is nearly as fast as register access. Here, for convenient reference as various routines are presented, is a complete list of all such memory:

```
__constant__ int d_ncases;          // Number of cases in complete training set
__constant__ int d_img_rows;         // Number of rows in input image
__constant__ int d_img_cols;         // Number of cols in input image
__constant__ int d_img_bands;        // Number of bands in input image
__constant__ int d_n_pred;           // Number of predictors
__constant__ int d_n_classes;        // Number of classes
__constant__ int d_n_classes_cols;   // Ditto, extended to multiple of 128 bytes (32 floats)
__constant__ int d_n_layers;         // Number of hidden layers
__constant__ int d_n_weights;        // Total number of weights across all layers
__constant__ int d_convgrad_cols[MAX_LAYERS]; // n_prior_weights[ilayer]
                                     // bumped up to multiple of 32
__constant__ int d_max_convgrad_each;   // Max hid * convwts_cols
                                     // in a CONV hid grad launch (work area per case)
__constant__ int d_layer_type[MAX_LAYERS];      // Type of each layer
__constant__ int d_nhid[MAX_LAYERS]; // N of neurons in each of the hidden layers
__constant__ int d_nhid_cols[MAX_LAYERS];       // Extended to mult of 128 bytes
__constant__ int d_height[MAX_LAYERS];          // Height (rows) of each layer
__constant__ int d_width[MAX_LAYERS];           // And width
__constant__ int d_depth[MAX_LAYERS];           // And number of slices
__constant__ int d_depth_cols[MAX_LAYERS];      // Ditto, extended to multiple of 128
__constant__ int d_n_prior_weights[MAX_LAYERS]; // N of inputs per neuron
```

```
__constant__ int d_HalfWidH[MAX_LAYERS];        // Horizontal half width
__constant__ int d_HalfWidV[MAX_LAYERS];        // And vertical
__constant__ int d_padH[MAX_LAYERS];            // Horizontal padding
__constant__ int d_padV[MAX_LAYERS];            // And vertical
__constant__ int d_strideH[MAX_LAYERS];         // Horizontal stride
__constant__ int d_strideV[MAX_LAYERS];         // And vertical
__constant__ int d_PoolWidH[MAX_LAYERS];        // Horizontal pooling width
__constant__ int d_PoolWidV[MAX_LAYERS];        // And vertical

static float *h_predictors = NULL;              // Training set; n_cases by n_pred
__constant__ float *d_predictors;

static int *h_class = NULL;                      // Class id is here
__constant__ int *d_class;

static double *activations = NULL;               // Activations of this layer
__constant__ double *d_act[MAX_LAYERS];          // Pointers to activation vector

static double *h_output = NULL;                  // Output activations
__constant__ double *d_output;

static int *h_poolmax_id[MAX_LAYERS];            // Used only for POOLMAX layer
__constant__ int *d_poolmax_id[MAX_LAYERS];      // Pointers to id vector each layer

static float *weights = NULL;                    // All weights, including output
__constant__ float *d_weights[MAX_LAYERS+1];     // Pointers to weight vector of each

static float *grad = NULL;                       // Gradient for all weights
__constant__ float *d_grad[MAX_LAYERS+1];        // Pointers to grad vector of each

static float *h_convgrad_work = NULL;            // Scratch for unflattened convolution grad
__constant__ float *d_convgrad_work;

static double *h_this_delta = NULL;              // Delta for current layer
__constant__ double *d_this_delta;

static double *h_prior_delta = NULL;             // Delta for next layer back
__constant__ double *d_prior_delta;

static float *h_ll_out = NULL;                   // Log likelihoods put here
__constant__ float *d_ll_out;
```

Initialization

Volumes 1 and 2 went into considerable detail in the "Initialization" section with the philosophy that because initialization is done first, it should appear first in the CUDA chapter. After some reflection, I decided to change this for Volume 3 and instead cover individual initialization topics in conjunction with the algorithms that rely on each topic. However, to illustrate an important general principle that appears repeatedly, this section examines the method for copying the training set from the host to the device.

   Cases in host memory are stored as doubles, but to save precious device memory, they are stored as floats on the device. Thus, we need to allocate scratch memory fdata to handle size translation. We also call cudaMalloc to allocate memory on the device. We transfer data from host memory to device memory in a set of nested loops that reorder it so that the band changes fastest. Finally, we copy the dataset to the device and copy the allocated pointer to d_predictors in the device's constant memory.

```
fdata = (float *) malloc (n_cases * n_pred * sizeof(float));
memsize = n_cases * n_pred * sizeof(float);        // Size of training set
error_id = cudaMalloc ((void **) &h_predictors, (size_t) memsize);

j = 0;
for (i=0; i<n_cases; i++) {                        // Move cases one at a time
  xptr = data + i * ncols;                         // Point to this case
  for (irow=0; irow<n_img_rows; irow++) {
    for (icol=0; icol<n_img_cols; icol++) {
      for (iband=0; iband<n_img_bands; iband++) // Band changes fastest on device
        fdata[j++] = (float) xptr[(iband*n_img_rows+irow)*n_img_cols+icol];
      }
    }
  }

error_id = cudaMemcpy (h_predictors, fdata, n_cases * n_pred * sizeof(float),
                    cudaMemcpyHostToDevice);
free (fdata);    // We no longer need this scratch memory
error_id = cudaMemcpyToSymbol (d_predictors, &h_predictors, sizeof(float *), 0,
                    cudaMemcpyHostToDevice);
```

Copying Weights to the Device

The initialization routine, called once after the architecture is set but before any computation is performed, allocates float memory on the device and fills in the pointer array that identifies the start of the weights for each layer. The first step in this allocation is to tally the total number of weights. Note that nhid_cols[ilayer] is the number of hidden neurons in this layer, bumped up to a multiple of 128 bytes (32 floats). The number of classes and the depth of convolutional layers are similarly bumped up. My convention is to append the suffix _cols to a quantity to indicate that the root quantity has been increased this way. The formula for bumping to a multiple of 32 is simple. In Equation 3-1, the division is integer division, discarding any remainder.

$$N_{bumped} = (N + 31) / 32 * 32 \qquad\qquad (3\text{-}1)$$

Here is the code that sums the number of weights and allocates sufficient memory on the device:

```
n_weights_on_device = 0;        // Counts total number of weights
for (ilayer=0; ilayer<= n_layers; ilayer++) { // For each of the hidden layers, plus final

  if (ilayer == n_layers)           // Output layer?
    n_weights_on_device += n_classes_cols * n_prior_weights[ilayer];

  else if (layer_type[ilayer] == TYPE_FC || layer_type[ilayer] == TYPE_LOCAL)
    n_weights_on_device += nhid_cols[ilayer] * n_prior_weights[ilayer];

  else if (layer_type[ilayer] == TYPE_CONV)
    n_weights_on_device += depth_cols[ilayer] * n_prior_weights[ilayer];

  else if (layer_type[i] == TYPE_POOLAVG || layer_type[i] == TYPE_POOLMAX)
    n_weights_on_device += 0;       // Just for clarity; pooling has no trainable weights

  } // For ilayer

memsize = n_weights_on_device * sizeof(float);
error_id = cudaMalloc ((void **) &weights, (size_t) memsize);
```

We now have to repeat the same sort of loop to fill in the pointer array that holds the starting address of the weights for each layer. Once this array of pointers is filled in, we copy it to constant memory on the device.

```
float *fptr[MAX_LAYERS+1];

n_total = 0;
for (ilayer=0;; ilayer++) {              // For each of the hidden layers, plus the output
  fptr[ilayer] = weights + n_total;   // Point to the weights for this layer

  if (ilayer >= n_layers)               // Do it through the output layer
    break;

  if (layer_type[ilayer] == TYPE_FC || layer_type[ilayer] == TYPE_LOCAL)
    n_total += nhid_cols[ilayer] * n_prior_weights[ilayer];

  else if (layer_type[ilayer] == TYPE_CONV)
    n_total += depth_cols[ilayer] * n_prior_weights[ilayer];

  else if (layer_type[i] == TYPE_POOLAVG || layer_type[i] == TYPE_POOLMAX)
    n_total += 0;                        // Just for clarity; pooling has no trainable weights

  } // For ilayer

error_id = cudaMemcpyToSymbol (d_weights, &fptr[0], (n_layers+1) * sizeof(float *),
                                    0, cudaMemcpyHostToDevice);
```

The code just shown is executed once, during initialization. But every time the weights change during the training process, we must recopy them to the device. This code is nasty because the weights are laid out on the host as shown on pages 31 (locally connected layers) and 34 (convolutional layers), but on the device they are laid out as shown on page 68, a very different ordering. The code for copying the weights to the device, properly ordered, is as shown now. Please study this code carefully to understand the weight layout because this will be important later when activation and backpropagation are shown.

```
int cuda_weights_to_device (
  int n_classes,          // Number of outputs
  int n_layers,           // Hidden layers; does not include output
  int *layer_type,        // Each entry (input to final) type
  int img_rows,           // Size of input image
  int img_cols,
  int img_bands,
```

```
  int *height,               // Height of visible field in each layer
  int *width,                // Width of visible field in each layer
  int *depth,                // Number of slices in each layer
  int *nhid,                 // Number of hidden neurons in each layer
  int *hwH,                  // Half-width of filters
  int *hwV,
  double **host_weights) // Vector of pointers to weights for each layer
{
  int n, n_prior, ilayer, ineuron, isub, n_cols_each;
  int idepth, iheight, iwidth, ndepth, nheight, nwidth;
  int in_row, in_col, in_slice, in_n_height, in_n_width, in_n_depth;
  double *wptr;
  float *fptr;
  cudaError_t error_id;

  fptr = fdata;              // Device weights will go here; fdata is already allocated

  for (ilayer=0; ilayer<=n_layers; ilayer++) {        // Process each layer individually
    wptr = host_weights[ilayer];                      // Host weights for this layer
/*
  Fully connected (output layer is always fully connected)
*/
    if (ilayer == n_layers || layer_type[ilayer] == TYPE_FC) {
      if (ilayer == 0) {
        in_n_height = img_rows;         // Size of layer feeding this layer
        in_n_width = img_cols;          // First hidden layer is fed by inputs
        in_n_depth = img_bands;
        }
      else {
        in_n_height = height[ilayer-1];    // Subsequent hidden layer is fed by prior
        in_n_width = width[ilayer-1];
        in_n_depth = depth[ilayer-1];
        }
```

```
n_prior = in_n_height * in_n_width * in_n_depth + 1; // N of weights per neuron

if (ilayer == n_layers)                    // Output layer?
  n = n_classes;                           // Equals depth in fully connected
else
  n = nhid[ilayer];                        // Equals depth in fully connected

n_cols_each = (n + 31) / 32 * 32; // For memory alignment to 128 bytes

for (in_row=0; in_row<in_n_height; in_row++) { // See page 68 for layout
  for (in_col=0; in_col<in_n_width; in_col++) {
    for (in_slice=0; in_slice<in_n_depth; in_slice++) {

      for (idepth=0; idepth<n; idepth++) { // Height and width are 1 in FC layer

        // Compute location of this neuron's weight vector in host
        isub = idepth*n_prior + (in_slice*in_n_height + in_row)*in_n_width + in_col;
        *fptr++ = (float) wptr[isub];
        } // For idepth

      while (idepth++ < n_cols_each) // Pad to multiple of 128 bytes
        *fptr++ = 0.0f;

      } // For in_slice
    } // For in_col
  } // For in_row

// Bias
for (idepth=0; idepth<n; idepth++) {

  // Compute location of this neuron's bias in host
  isub = idepth * n_prior + n_prior - 1;
  *fptr++ = (float) wptr[isub];
  } // For idepth

while (idepth++ < n_cols_each) // Pad to multiple of 128 bytes
  *fptr++ = 0.0f;
  }

/*

  Locally connected layer

*/
```

```
else if (layer_type[ilayer] == TYPE_LOCAL) {
  // For LOCAL layers, neuron and filter layout is (height, width, depth).
  n = nhid[ilayer];
  n_cols_each = (n + 31) / 32 * 32;       // For memory alignment to 128 bytes

  ndepth = depth[ilayer];                 // Size of the current layer
  nheight = height[ilayer];
  nwidth = width[ilayer];

  in_n_height = 2 * hwV[ilayer] + 1;      // Filter rectangle dimensions
  in_n_width = 2 * hwH[ilayer] + 1;
  if (ilayer == 0)                        // First hidden layer
    in_n_depth = img_bands;               // so input in image
  else                                    // Subsequent hidden layer
    in_n_depth = depth[ilayer-1];         // Fed by prior hidden layer

  n_prior = in_n_height * in_n_width * in_n_depth + 1;     // N weights per neuron

  for (in_row=0; in_row<in_n_height; in_row++) {           // See page 68 for layout
    for (in_col=0; in_col<in_n_width; in_col++) {
      for (in_slice=0; in_slice<in_n_depth; in_slice++) {

        for (iheight=0; iheight<nheight; iheight++) { // nhid = ndepth*nheight*nwidth
          for (iwidth=0; iwidth<nwidth; iwidth++) {
            for (idepth=0; idepth<ndepth; idepth++) {

              // Compute location of this neuron's weight in host
              // We do this in two steps.
              // First, locate the neuron in the current layer.
              // Multiply this by the number of weights per current neuron (n_prior)
              // Then add the location in the filter rectangle

              isub = (idepth * nheight + iheight) * nwidth + iwidth; // Current layer loc
              isub = isub*n_prior+(in_slice*in_n_height+in_row)*in_n_width+in_col;
              *fptr++ = (float) wptr[isub];
              } // For idepth
            } // For iwidth
          } // For iheight
        // The entire current layer for this single input location is done. Pad.
        ineuron = nhid[ilayer];
```

```
          while (ineuron++ < n_cols_each) // Pad to multiple of 128 bytes
            *fptr++ = 0.0f;

          } // For in_slice
        } // For in_col
      } // For in_row

    // Bias
    for (iheight=0; iheight<nheight; iheight++) { // nhid = ndepth * nheight * nwidth
      for (iwidth=0; iwidth<nwidth; iwidth++) {
        for (idepth=0; idepth<ndepth; idepth++) {

          // Compute location of this neuron's weight vector in host
          isub = (idepth * nheight + iheight) * nwidth + iwidth; // Neuron in this layer
          isub = isub * n_prior + n_prior - 1; // Location of bias
          *fptr++ = (float) wptr[isub];

          } // For idepth
        } // For iwidth
      } // For iheight

    // Pad the bias set
    ineuron = nhid[ilayer];
    while (ineuron++ < n_cols_each) // Pad to multiple of 128 bytes
      *fptr++ = 0.0f;
    }
/*
  Convolutional layer
*/

    else if (layer_type[ilayer] == TYPE_CONV) {

    nheight = height[ilayer];                    // Size of the current layer
    nwidth = width[ilayer];
    ndepth = depth[ilayer];

    n_cols_each = (ndepth + 31) / 32 * 32;  // For memory alignment to 128 bytes
    in_n_height = 2 * hwV[ilayer] + 1;       // Size of the filter rectangle
    in_n_width = 2 * hwH[ilayer] + 1;
    if (ilayer == 0)
      in_n_depth = img_bands;
```

```
    else
      in_n_depth = depth[ilayer-1];

    n_prior = in_n_height * in_n_width * in_n_depth + 1;   // N of weights per neuron

    for (in_row=0; in_row<in_n_height; in_row++) {         // See page 68 for layout
      for (in_col=0; in_col<in_n_width; in_col++) {
        for (in_slice=0; in_slice<in_n_depth; in_slice++) {
          for (idepth=0; idepth<ndepth; idepth++) {

            // Compute location of this neuron's weight vector in host
            isub = idepth*n_prior + (in_slice*in_n_height + in_row)*in_n_width + in_col;
            *fptr++ = (float) wptr[isub];
            } // For idepth

          // All current-layer depths for this filter element are done. Pad.
          while (idepth++ < n_cols_each) // Pad to multiple of 128 bytes
            *fptr++ = 0.0f;
          } // For in_slice
        } // For in_col
      } // For in_row

    // Bias
    for (idepth=0; idepth<ndepth; idepth++) {
      // Compute location of this neuron's bias in host
      isub = idepth * n_prior + n_prior - 1;
      *fptr++ = (float) wptr[isub];
      } // For idepth

    // Pad the bias
    while (idepth++ < n_cols_each) // Pad to multiple of 128 bytes
      *fptr++ = 0.0f;
    }
  } // For ilayer
  error_id = cudaMemcpy (weights, fdata, n_weights_on_device * sizeof(float),
                          cudaMemcpyHostToDevice);
  return 0;
}
```

Activating the Output Layer

We'll ease into the CUDA code with the simplest routine. The code shown here is for the usual situation of the model containing at least one hidden layer. The routine for the situation of no hidden layer can be found in MOD_CUDA.cu but will not be listed here, as it is practically identical to this and offers no new insights.

Here is the host routine that is called from the supervisor routine. This will often be a bit inefficient because the number of classes will usually be less than the warp size (32), resulting in incomplete warps, generally a severe no-no. However, the fraction of actual runtime taken by this step is almost invisibly tiny, so trading some inefficiency for simplicity is good. The limitation of block size to four warps is arbitrary but reasonable; feel free to change it if you want.

```
int cuda_output_activation (
  int istart,    // First case in this batch
  int istop      // One past last case
  )
{
  int warpsize, threads_per_block;
  dim3 block_launch;
  cudaError_t error_id;

  warpsize = deviceProp.warpSize;    // Threads per warp, likely 32 well into the future

  threads_per_block = (n_classes + warpsize - 1) / warpsize * warpsize;
  if (threads_per_block > 4 * warpsize)       // This is arbitrary but reasonable
    threads_per_block = 4 * warpsize;

  block_launch.x = (n_classes + threads_per_block - 1) / threads_per_block;
  block_launch.y = istop - istart;
  block_launch.z = 1;

  device_output_activation <<< block_launch, threads_per_block >>> (istart);
  cudaDeviceSynchronize();
  return 0;
}
```

The device code for performing this task is shown on the next page. The following issues should be noted:

- The computed outputs for the *entire* training set are saved in device memory. This facilitates rapid criterion computation later, and it prepares the way for more advanced performance stats as well as the dumping of all results to the host if desired.

- Intermediate results such as activations are retained *only for the batch* currently being processed. This saves valuable device memory.

- The implication of these two facts is that we need the batch start, istart, to offset the output storage properly, but istart is not used when referencing activations.

- The most time-critical line in this code is the sum += *wptr * inptr[i_input] line. This has two global accesses, and it is inside a loop.

- The reference to inptr[i_input] in this line is independent of the thread index, which makes it unavoidably impossible to coalesce. But for this same reason, it has the same value for all threads, and hence a single read will service all threads simultaneously, which is very efficient.

- The reference to the weight is perfectly coalesced because weights are ordered with the output neuron changing fastest, which is defined by the thread. Moreover, it is padded so that each warp starts on a 128-byte address.

- The output storage, while not 128-byte aligned (that would waste too much memory), is nevertheless coalesced in that adjacent threads write to adjacent memory.

```
__global__ void device_output_activation (
  int istart       // First case in this batch; needed for output
  )
{
  int icase, iout, i_input, n_inputs;
  double sum;
  float *wptr;
  double *inptr;
```

```
iout = blockIdx.x * blockDim.x + threadIdx.x;

if (iout >= d_n_classes)
   return;

icase = blockIdx.y;                      // Activities are zero origin, not offset by istart

wptr = d_weights[d_n_layers] + iout; // Current neuron weight changes fastest

n_inputs = d_nhid[d_n_layers-1];
inptr = d_act[d_n_layers-1] + icase * n_inputs;  // Feed from prior layer is here

sum = 0.0;                               // Will cumulate logit
for (i_input=0; i_input<n_inputs; i_input++) {   // Equation 1-9 on page 16
   sum += *wptr * inptr[i_input];
   wptr += d_n_classes_cols;            // Weights are zero-padded to 128 bytes
   }

sum += *wptr;                           // Bias

d_output[(icase+istart)*d_n_classes+iout] = sum;   // We save the logit
}
```

Activating Locally Connected and Convolutional Layers

This is the first of two CUDA routines for computing the activation of locally connected and convolutional layers. It is the easier of the two to understand and is a prerequisite to understanding the second. The second algorithm uses shared memory to speed operation. Nevertheless, the routine presented in this section is necessary, as it handles "cleanup" operations that will be discussed later. So, studying this code is far from a waste of time.

```
int cuda_hidden_activation_LOCAL_CONV (
   int local_vs_conv,        // Is this a LOCAL (vs CONV) layer?
   int istart,               // First case in this batch
   int istop,                // One past last case
```

```
  int nhid,               // Number of hidden neurons in this layer = H*W*D
  int n_slices,           // Depth of this layer
  int ilayer              // Layer to process
  )
{
  int warpsize, threads_per_block;
  dim3 block_launch;
  cudaError_t error_id;

  warpsize = deviceProp.warpSize;      // Threads per warp, likely 32 well into the future

  threads_per_block = (n_slices + warpsize - 1) / warpsize * warpsize;
  if (threads_per_block > 4 * warpsize)
    threads_per_block = 4 * warpsize;        // Arbitrary but reasonable

  block_launch.x = (n_slices + threads_per_block - 1) / threads_per_block;
  block_launch.y = nhid / n_slices;  // Visual field size; MUST be less than 65535!
  block_launch.z = istop - istart;      // Number of cases in this batch

  device_hidden_activation_LOCAL_CONV <<< block_launch, threads_per_block >>>
                      (local_vs_conv, istart, 0, 0, n_slices, ilayer);

  cudaDeviceSynchronize();
  return 0;
}
```

The device code is quite long and complex, so I'll present it in sections, with explanations of operation interspersed. We begin with the calling parameter list. Several of the parameters might be a little confusing at this point. We are already familiar with case_start; this is just istart in the interface routine, the case in the training set that begins the batch currently being processed. Breaking the training set into batches has the important purpose of limiting the time taken by each launch so we can avoid the infamous Windows WDDM timeout error. It also facilitates more advanced operations, such as cross validation or walkforward testing.

The next two parameters, case_offset and slice_start, are specialized. The former applies an additional offset to the case being processed in the batch, and the latter lets us begin processing with slices past the first. If this routine were being exclusively used for computing the activation, both of these would be zero. However, we will see later that these offsets are needed when this routine is used for cleanup after the shared-memory version.

Each depth slice is handled by a thread, as shown here. We then compute the location in the current layer of the neuron being activated.

```
__global__ void device_hidden_activation_LOCAL_CONV (
  int local_vs_conv,      // Is this a LOCAL (vs CONV) layer?
  int case_start,         // First case in this batch (relative to dataset)
  int case_offset,        // Offset relative to this batch (used in shared version)
  int slice_start,        // First slice in this batch
  int n_slices,           // Number of slices to be done in this launch
  int ilayer              // Layer to process
  )
{
  int kwt, kin, wtsub, insub, iheight, iwidth, idepth, n_height, n_width, n_depth, wt_c ols;
  int rstart, rstop, cstart, cstop, rbase, cbase, in_slice, in_row, in_col, ihid, nH;
  float *f_inptr, *wptr;
  double sum, *actptr;

  idepth = blockIdx.x * blockDim.x + threadIdx.x;
  if (idepth >= n_slices)
    return;
  idepth += slice_start;
  iheight = blockIdx.y / d_width[ilayer];
  iwidth = blockIdx.y % d_width[ilayer];

  nH = 2 * d_HalfWidH[ilayer] + 1;    // We'll reference this deep inside a loop later
```

We are about to compute the activation of neuron (iheight, iwidth, idepth) in this layer. Note that it is critical that idepth be associated with the thread. This ensures that adjacent threads reference the same input, which allows efficient memory use (a single global fetch services all threads in the warp). Also, the weights are ordered so that depth-fastest changes produce perfect or very good coalescing. Thus, the neuron layout in the current layer is (height, width, depth).

This layout gives strong motivation for locally connected layers to have the depth be a multiple of 32. To see why, note the ihid= line in the following code. That multiplication ensures perfect, as opposed to just very good, coalescing of the weight fetches (as long as slice_start is zero; if not, the coalescing is still very good).

We note with a comment that the case (not yet offset) is in the *z* dimension of the block. We fetch the padded length of each row of the weight matrix and find the location in the weight array of the first filter weight. Just to pound it in, observe that adjacent threads reference adjacent weights.

```
// icase = blockIdx.z; // Avoid using a register by directly referencing it later

if (local_vs_conv) {
  wt_cols = d_nhid_cols[ilayer]; // Padded size of weight matrix rows
  ihid = (iheight * d_width[ilayer] + iwidth) * d_depth[ilayer] + idepth;
  wptr = d_weights[ilayer] + ihid;
  }

else {
  wt_cols = d_depth_cols[ilayer];
  wptr = d_weights[ilayer] + idepth; // First filter weight for this slice is here
  }
```

Just as was done on page 32 when activation of locally connected and convolutional layers was first discussed (please review that if needed), we compute the bounds of the rectangle in the prior layer, which contributes to the activation of the current neuron. We keep start/stop bounds, which do not extend over the boundaries of the prior layer's visual field, and we also keep "base" bounds, which let us locate positions in the filter rectangle.

```
sum = 0.0;

// Center of first filter is at HalfWidth-Pad; filter begins at -Pad.
rbase = rstart = d_strideV[ilayer] * iheight - d_padV[ilayer];
rstop = rstart + 2 * d_HalfWidV[ilayer];
cbase = cstart = d_strideH[ilayer] * iwidth - d_padH[ilayer];
cstop = cstart + 2 * d_HalfWidH[ilayer];

if (rstart < 0)     // These limit the top and left
  rstart = 0;       // We'll limit the bottom and right below
if (cstart < 0)
  cstart = 0;
```

We must duplicate the same code for the situation of this being the first hidden layer (fed by the input) versus a subsequent hidden layer (fed by prior activations). The input uses a float pointer, and activations a double pointer. Deciding which pointer to use in the inner loop would be too slow!

If this is the first hidden layer, get a pointer to the input case, taking both case offsets into account. Also, limit the bottom and right rectangle bounds to not extend past the input's visual field.

```
if (ilayer == 0) {

  f_inptr = d_predictors + (blockIdx.z + case_offset + case_start) * d_n_pred;

  if (rstop >= d_img_rows)
    rstop = d_img_rows - 1;
  if (cstop >= d_img_cols)
    cstop = d_img_cols - 1;
```

Sum the filter over the prior layer's rectangle for all prior-layer slices. By using the start/stop limits, we avoid the inefficient check for being outside the visual field that was used in the code we saw in MOD_NO_THR.CPP.

The indexing inside this loop may be a bit confusing. We compute (in_row - rbase) * nH + in_col - cbase as the position occupied by (in_row, in_col) in the visual field of the filter. If this is not clear, draw a rectangle of dots representing the filter elements and confirm this math. One of these filter rectangles exists for each input slice, but the filter is ordered with the slice changing fastest. So we multiply this filter visual field position by the number of slices (bands for the input) and then add the slice to get the exact filter element. If this latter operation is not clear, make a small stack of your dotted rectangles and realize that counting moves up the stack before changing position in the visual field. Similar math locates the input. For extra clarity, two commented lines show the math that's really going on. Last of all, we add in the bias.

```
for (in_row=rstart; in_row<=rstop; in_row++) {
  kwt = (in_row - rbase) * nH;
  kin = in_row*d_img_cols;

  for (in_col=cstart; in_col<=cstop; in_col++) {
    wtsub = (kwt + in_col - cbase) * d_img_bands;
    insub = (kin+in_col) * d_img_bands;
```

```
      for (in_slice=0; in_slice<d_img_bands; in_slice++) {
        // wtsub = ((in_row - rbase) * nH + in_col - cbase) * d_img_bands + in_slice;
        // insub = (in_row*d_img_cols+in_col)*d_img_bands+in_slice;
        sum += f_inptr[insub] * wptr[wtsub*wt_cols];
        ++wtsub;
        ++insub;
        } // For in_slice
      } // For in_col
    } // For in_row

  sum += wptr[(d_n_prior_weights[ilayer]-1) * wt_cols];     // Bias
  }
```

If this is a subsequent hidden layer, rather than the first, the operation is nearly identical to what we just saw for the first hidden layer. The only difference is that we now reference the prior hidden layer rather than the input image.

```
else {
  actptr = d_act[ilayer-1] + (blockIdx.z + case_offset) * d_nhid[ilayer-1];
  n_height = d_height[ilayer-1];    // Size of the layer feeding this one
  n_width = d_width[ilayer-1];
  n_depth = d_depth[ilayer-1];

  if (rstop >= n_height)                 // Don't go outside prior layer's visual field
    rstop = n_height - 1;
  if (cstop >= n_width)
    cstop = n_width - 1;

  for (in_row=rstart; in_row<=rstop; in_row++) {
    kwt = (in_row - rbase) * nH;
    kin = in_row*n_width;

    for (in_col=cstart; in_col<=cstop; in_col++) {
      wtsub = (kwt + in_col - cbase) * n_depth;
      insub = (kin+in_col) * n_depth;

      for (in_slice=0; in_slice<d_depth[ilayer-1]; in_slice++) {
        // This is what we are really doing
        // wtsub = ((in_row - rbase) * nH + in_col - cbase) * n_depth + in_slice;
        // insub = (in_row*n_width+in_col)*n_depth+in_slice;
```

```
        sum += actptr[insub] * wptr[wtsub*wt_cols];
        ++wtsub;
        ++insub;
        } // For in_slice
      } // For in_col
    } // For in_row

  sum += wptr[(d_n_prior_weights[ilayer]-1) * wt_cols];     // Bias
  }
```

Before leaving that section of the code, it is worth noting several vital facts.

- There are two global reads, and they happen in the innermost of a set of nested loops, so they are critical.

- The read of the input value is independent of the thread (idepth), which means that it cannot be coalesced. But for the same reason, it has the same value for all threads in the warp, so a single fetch services all threads with an efficient mass broadcast.

- The other global read is the filter weight. This is offset by the thread (idepth), so adjacent threads access adjacent memory locations, resulting in very good coalescing. Moreover, if slice_start is zero, all warps begin at a multiple of 128 bytes (note the multiplication by wt_cols, which is a multiple of 32), resulting in perfect coalescing.

Finally, we apply the hyperbolic tangent activation function and store the computed activation. Note that ihid varies with idepth so that adjacent threads write to adjacent memory locations, resulting in very good coalescing. As a bonus, if the depth of the current layer is a multiple of 32, and if slice_start is zero, coalescing will be perfect.

```
if (sum > MAX_EXP)
  sum = 1.0;
else {
  sum = exp (2.0 * sum);
  sum = (sum - 1.0) / (sum + 1.0);
  }

n_height = d_height[ilayer];
n_width = d_width[ilayer];
n_depth = d_depth[ilayer];
```

```
  actptr = d_act[ilayer];
  ihid = (iheight * n_width + iwidth) * n_depth + idepth;
  actptr[(blockIdx.z+case_offset)*d_nhid[ilayer]+ihid] = sum;
}
```

Using Shared Memory to Speed Computation

This section presents a method for significantly speeding computation of activations. Be warned that this topic is considerably more complex than the algorithm shown in the prior section, and understanding it will be hopeless unless the prior algorithm is thoroughly understood.

The underlying basis of the algorithm shown here is that shared memory has tremendously faster access than global memory. The algorithm of the prior section repeatedly fetches the same global memory, resulting in much redundancy. In truth, the penalty is not terribly severe because I took enormous pains to ensure that all global memory accesses are as fast as possible by ensuring very good or perfect coalescing everywhere. Moreover, computations are structured in such a way that mathematical operations effectively hide much of memory fetching stalls. For many or most applications, the mathematics pipeline is the dominant limiting factor. Still, clean, modern CUDA programming demands that we take advantage of fast shared memory whenever feasible.

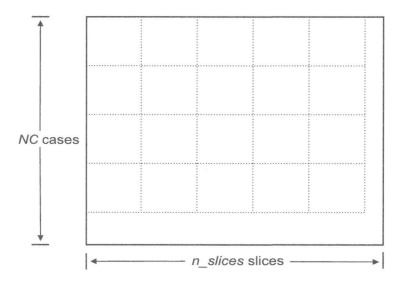

Figure 3-1. *Block layout for activation with shared memory*

Figure 3-1 illustrates what we will be doing. Imagine a grid in which the vertical dimension represents every case in the set we will be processing, and the horizontal dimension represents every slice in the layer being computed. The (row, column) position in the visual field is irrelevant to this discussion; it is already specified as a fixed location. Thus, any spot on the grid in this figure represents the activation of a single neuron with a prespecified position in the visual field, and slice according to its horizontal location in the figure, and for a case represented by the vertical location on the figure.

How would we compute this activation? Imagine that we stack a bunch of these figures, overlaid on top of one another. Each layer in this stack (not to be confused with layers in the model!) represents a position in the visual field of the prior-layer rectangle (which does not concern us now), as well as a slice within the prior rectangle. Two key numbers are associated with this position: an activation in the prior layer and the corresponding filter weight. And one more layer of this figure will represent the bias term.

Thus, to compute the activation of the single neuron under discussion, we look at the spire of elements coming up and out of the page, all emanating from a single point in Figure 3-1. Cumulate the dot product associated with this spire. There will be n_prior_ weights in this spire.

Notice that Figure 3-1 is subdivided into squares delineated with dotted lines. Each such square represents a single launch block. All of these blocks will be computed in a single kernel launch. The size (length and width) of these blocks should be as large as possible for maximum efficiency, subject to the constraint that the square of this length/width must not exceed the hardware limit on the number of threads per block. My CONVNET program uses 32 because modern devices have a limit of at least 1,024 threads per block. The number of global memory fetches is reduced by roughly a factor equal to the length/width of the block, which can be substantial.

For the moment, ignore the extra rows and columns outside an integral number of blocks. We'll deal with these later, as a last step.

Before continuing, let's take a quick look at the launch parameters. The launched blocks will be big, 32*32=1024 threads here. The current-layer slices will be divided into blocks along the x dimension, and the cases in this batch will be divided along the y dimension. The z dimension will specify the (row, column) location in the visible field of the current layer.

```
dim3 thread_launch, block_launch;

nc = istop - istart;        // Number of cases

thread_launch.x = BLOCK_SIZE;        // 32 here
thread_launch.y = BLOCK_SIZE;        // These must be the same
thread_launch.z = 1;

block_launch.x = n_slices / BLOCK_SIZE; // Number of blocks horizontally
block_launch.y = nc / BLOCK_SIZE;       // And vertically
block_launch.z = nhid / n_slices;       // Height times width; visual field size

device_hidden_activation_LOCAL_CONV_shared
            <<< block_launch, thread_launch >>> (local_vs_conv, istart, ilayer);
```

Let's look at a rough overview of the device algorithm. Each block in the launch will completely compute the activations of all neurons/cases in the block shown enclosed in dotted lines in Figure 3-1. In addition to being structured in BLOCK_SIZE squares in the slice and case dimensions, the "up from the page" direction, which represents prior-layer activations and filter weights, will also be processed in BLOCK_SIZE chunks. So we are actually dealing with cubes. With this in mind, here are the steps that we will soon examine in detail:

1. Do all preliminary calculations. Get pointers to the filter weights and the activations that are feeding the current layer, whether these be from the input image or from a prior hidden layer. Find the bounds of the prior-layer rectangle over which the filter acts.

2. Get the number of elements that will go into the dot product of activations and filter weights. This will be n_prior_weights in the interior and less at the borders if padded. Call this n_inner.

3. inner_blocks = (n_inner + BLOCK_SIZE - 1) / BLOCK_SIZE is the number of "inner-loop" blocks that will be needed to sum the dot product emanating up and out of the page, each inner-loop block handling BLOCK_SIZE terms in the dot product.

4. Perform the following computation loop, in which s_slices and s_cases are shared memory matrices BLOCK_SIZE square.

```
sum = 0;
for (inner=0; inner<inner_blocks; inner++) {
```

Slice is derived from threadIdx.x
Inner index is derived from threadIdx.y

```
    s_slices[threadIdx.y][threadIdx.x] = weight [ inner index, slice ];
```

Case is derived from threadIdx.y
Inner index is derived from threadIdx.x

```
    s_cases[threadIdx.y][threadIdx.x] = activation [ case, inner index ];
```

Wait for all threads to complete the above global fetches.

```
    for (k=0; k<BLOCK_SIZE; k++)
        sum += s_cases[threadIdx.y][k] * s_slices[k][threadIdx.x];
```

Wait for all threads to complete the above summation loop.

```
} // End of 'inner_blocks' loop
```

5. Apply hyperbolic tangent activation function and save result.

It is absolutely *crucial* that you understand the algorithm just shown. Without a solid understanding of that little algorithm, you won't have a chance of understanding the code. So let's walk through it. To simplify the discussion, we will assume that BLOCK_SIZE is 32.

We are executing a block of 32*32=1024 threads. The task of this block is to compute the activations for a given fixed (z block dimension) position in the current layer's visual field and for a set of 32 depths and 32 cases. Ignore the z dimension, the position of the current neuron in the visible field. It has no bearing on the algorithm under discussion, and thoughts of it will just confuse things. Remember only that we are computing 1,024 activations in this block, neurons at 32 slices for each of 32 cases.

The loop shown earlier loops through sets of 32 filter weight/prior-layer activation pairs. In other words, the dot product for computing the activation is evaluated in chunks of 32 pairs at a time, one such 32-pair set for each pass through the outer loop. Thus, the dot product will not be completed until all passes through the outer loop are completed. This dot product will be cumulated in sum.

The first step in the loop is for the 1,024 threads to cooperatively fetch from global memory the filter weights for this current-layer slice in the filter. Recall that locally connected and convolutional layers have a different set of filter weights for each slice. Note that the slice of the weight is derived from the x index of the thread, so we have very good or perfect coalescing.

The next step is for the 1,024 threads to cooperatively fetch the other item in each pair, the activation of 32 prior-layer neurons for 32 cases. If you trace in the code the evolution of the subscript for the activation, you'll see that it derives from the x dimension of the thread, meaning once again that this global fetch is very well or perfectly coalesced.

At this point, the block has to pause as necessary to wait for all warps to finish these two tasks. Remember that the warp scheduler does not guarantee perfect coordination among warps. We must not continue until all of these quantities have been fetched into shared memory.

The last step is to sum this inner-loop block's 32 components of the dot product. Each pass through that inner loop has two accesses that would otherwise be global but that now can use the shared memory. This lets us get the redundant fetches from fast shared memory instead of slow global memory.

When this algorithm is complete for a thread, sum contains the complete dot product for a neuron in one of the 32 slices for one of the 32 cases. This is an entry in the layout shown in Figure 3-1.

It's worthwhile to do a quick comparative performance analysis of this algorithm. To keep things simple, assume inner_blocks=1, so we are concerned with a single pass through the loop. The analysis to come applies regardless of how many passes are executed.

Because there are 32*32=1024 threads, the first and second steps each do 1,024 global memory loads. So at that point we have 2,048 slow loads. Now look at the summation loop. The single sum line is executed on 1,024 threads, with two loads each. The loop executes 32 times, so we have 64K loads. If we had not staged the values to shared memory first, we would be doing 64K slow loads. But because that loop accesses fast shared memory, we are burdened with just the 2K slow loads to initialize. We have gained by a factor of 32, the block size. Of course, there is a small amount of overhead involved, so the speedup is not quite that high, but it can be significant.

And to throw a little more cold water on this shared-memory approach, remember that the speedup applies only to global memory accesses. If "slow" global memory accesses are well coalesced and good programmers always make sure to do so, then other delays come into play as limiting factors. The math pipeline has finite capacity, and many serial operations rely on completion of prior steps, resulting in execution dependencies. So in truth, the speedup because of shared memory is often not nearly as substantial as might be hoped. Still, it is a worthwhile endeavor.

Device Code

That brief summary of the algorithm skirted many important issues, but it is the essence of the technique. Please don't go on until you are comfortable with your understanding of that outline. When you are ready, take a deep breath. Here we go. The calling parameter list and variable declarations are as follows:

```
__global__ void device_hidden_activation_LOCAL_CONV_shared (
  int local_vs_conv,        // Is this a LOCAL (vs CONV) layer?
  int istart,               // First case in this batch
  int ilayer                // Layer to process
  )
{
  int k, iheight, iwidth, idepth, icase, n_height, n_width, n_depth, wt_cols;
  int ihid, inner, n_inner, inner_blocks, prod;
  int rstart, rstop, cstart, cstop, rbase, cbase, in_slice, in_row, in_col, isub, nH;
  float *f_inptr, *wptr;
  double value, sum, *actptr;
```

In a block, threadIdx.x and threadIdx.y are the location within the BLOCK_SIZE square block. The entire matrix of cases (row) by slices (column) is divided into these blocks, each of which is a launched block whose location in the entire matrix is given by blockIdx.x and blockIdx.y. The sharing logic ignores blockIdx.z, which is just the location in the visual field. The next four quantities identify the location within the entire matrix, and nH is the horizontal dimension of the filter.

```
idepth = blockIdx.x * BLOCK_SIZE + threadIdx.x;   // Slice in current layer
icase = blockIdx.y * BLOCK_SIZE + threadIdx.y;    // Offset of case in this batch

iheight = blockIdx.z / d_width[ilayer];           // Row in visual field
iwidth = blockIdx.z % d_width[ilayer];            // And column

nH = 2 * d_HalfWidH[ilayer] + 1;                  // Horizontal width of the filter
```

This thread will compute the activation of neuron (iheight, iwidth, idepth) for case icase. These four quantities were just computed from the block and thread indices. We now get a pointer to the filter weights for this neuron. Note that it is critical that idepth be associated with threadIdx.x, for several reasons. Recall that weights are zero padded and ordered so that depth changes fastest. Having threads also change with depth ensures perfect coalescing of weights. Also, the neuron layout in a layer is (row, column, slice). Thus, adjacent threads will have the same position in the visual field and hence reference the same input activation, meaning that the hardware can broadcast this single loaded value across the entire warp, resulting in extremely efficient activation fetches.

We also need wt_cols, the padded length of rows of the weight matrix. Locally connected layers will have *nhid* weights, followed by padding to bring the length up to a multiple of 32 floats (128 bytes). Convolutional layers will have *depth* weights, again padded to 128 bytes. Note in the following code that the multiplication by d_depth[ilayer] provides strong motivation for the user to let locally connected layers have a depth that is a multiple of 32. This makes the difference between very good versus perfect coalescing in locally connected layers. Convolutional layers are always perfectly coalesced.

```
if (local_vs_conv) {                      // Is this a locally connected layer?
    wt_cols = d_nhid_cols[ilayer];       // Padded size of weight matrix rows
    ihid = (iheight * d_width[ilayer] + iwidth) * d_depth[ilayer] + idepth;
    wptr = d_weights[ilayer] + ihid;
    }

else {
    wt_cols = d_depth_cols[ilayer]; // Padded size of weight matrix rows
    wptr = d_weights[ilayer] + idepth;
    }
```

That took care of finding pointers to the weights, which are one component of the dot-product pairs. The activations in the prior layer are the other component. First, we get a pointer to the prior-layer activations and the size of this prior layer.

```
if (ilayer == 0) {
    f_inptr = d_predictors + (icase + istart) * d_n_pred;
    n_height = d_img_rows;
    n_width = d_img_cols;
    n_depth = d_img_bands;
    }
```

```
else {
  actptr = d_act[ilayer-1] + icase * d_nhid[ilayer-1];
  n_height = d_height[ilayer-1];
  n_width = d_width[ilayer-1];
  n_depth = d_depth[ilayer-1];
  }
```

Now locate the rectangle in the prior layer that corresponds to the neuron being computed in the current layer. I'll repeat the short discussion that appeared earlier in the context of serial (non-CUDA) computation.

Computation of the activation of the current neuron is based on a rectangle in the *prior* layer whose position is determined by the position (iheight, iwidth) of the current neuron in the visual field of *this* layer. In both the vertical and horizontal directions, the center of the first filter (first row or column of the current layer) is at the location *HalfWidth-Pad* in the prior layer, and the first row/column of this first rectangle is at *-Pad*, which will be in the zero-padding area if padding is done. If this is not clear, please draw yourself a little one-dimensional picture.

This tells us how to compute the inclusive starting and stopping rows and columns of the rectangle in the prior layer, which contributes to the activation of the neuron in the current layer. We start at *-Pad*, advance by *Stride* as the current layer advances, and end at twice the *HalfWidth*. We need the *start/stop* values so we know if we are in a zero-padded edge, and we need the *base* values so we can locate our position in the filter rectangle.

```
rbase = rstart = d_strideV[ilayer] * iheight - d_padV[ilayer];
rstop = rstart + 2 * d_HalfWidV[ilayer];
cbase = cstart = d_strideH[ilayer] * iwidth - d_padH[ilayer];
cstop = cstart + 2 * d_HalfWidH[ilayer];

if (rstart < 0)
  rstart = 0;
if (cstart < 0)
  cstart = 0;

if (rstop >= n_height)
  rstop = n_height - 1;
if (cstop >= n_width)
  cstop = n_width - 1;
```

Everything is ready for gathering the two components of the dot product and saving them in fast shared memory. Turn back to page 90 and quickly review the general outline of this algorithm. We now perform step 3 and show the beginning of step 4. The rectangle over which the summation is taking place may include zero padding outside an edge of the prior layer, so we need to take this into account when we compute the number of components in the dot product.

We let prod be the number of elements in each row of this rectangle and then multiply this by the number of rows and add 1 (the bias term) to get n_inner, the total number of terms in the dot product that we will sum. This will be divided into inner_blocks blocks, which must include a possible incomplete block at the end.

```
prod = (cstop-cstart+1) * n_depth;     // Each prior-layer row has this many elements
n_inner = (rstop-rstart+1) * prod + 1; // This many terms in inner sum (+1 is for bias)
inner_blocks = (n_inner + BLOCK_SIZE - 1) / BLOCK_SIZE;

sum = 0.0;

for (inner=0; inner<inner_blocks; inner++) {
    __shared__ double s_cases[BLOCK_SIZE][BLOCK_SIZE];
    __shared__ float s_slices[BLOCK_SIZE][BLOCK_SIZE];
```

The serial version of this algorithm was relatively simple because it just summed over the triple-nested loop of rows, columns, and slices. The parallel version is a lot more complicated because each term in the sum is handled independently by a different thread. So for each term we must locate it in the prior-layer weight and activation volumes. That's fussy. We'll begin with the weights, showing the code first and explaining later.

```
isub = inner * BLOCK_SIZE + threadIdx.y;    // Ordinal position in dot product loop
if (isub >= n_inner)                // Outside inner block
    value = 0.0;                    // The last block is likely incomplete
else if (isub == n_inner-1)         // Bias
    value = wptr[(d_n_prior_weights[ilayer]-1) * wt_cols]; // Bias is last weight
else {
    in_row = isub / prod;
    k = isub - in_row * prod;
    in_col = k / n_depth;
    in_slice = k % n_depth;
```

```
      in_row += rstart;
      in_col += cstart;
      isub = ((in_row - rbase) * nH + in_col - cbase) * n_depth + in_slice;
      value = wptr[isub*wt_cols];
      }
   s_slices[threadIdx.y][threadIdx.x] = value;
```

We computed prod as the number of elements in each row of the dot product being summed. Divide the ordinal position by this to get the *relative* row in the rectangle, and remove this component from the position. Divide by the depth to get the relative column, and the remainder is the slice. Add the starting positions of the rectangle to get the *actual* positions in the prior-layer visual field.

If our rectangle extends over an edge into zero-padded territory, the coordinates of the rectangle in the context of the terms of the dot product summation will not correspond to those in the visual field, so to get the relative position in the filter rectangle we subtract the base to get the subscript in the filter weight set. We could save one operation in the row and one in the column by saving *start* minus *base* outside the loop, but I wrote it this way for clarity. Some readers may want to fix this.

Gathering the prior-layer activations is similar to what we just did for the weights. Here is the code, and I'll mention only the few significant differences:

```
isub = inner * BLOCK_SIZE + threadIdx.x;   // Ordinal position in dot product loop
if (isub >= n_inner)               // Outside inner block
   value = 0.0;                    // Last block is likely incomplete
else if (isub == n_inner-1)        // Bias
   value = 1.0;
else {
   in_row = isub / prod;
   k = isub - in_row * prod;
   in_col = k / n_depth;
   in_slice = k % n_depth;
   in_row += rstart;
   in_col += cstart;
   isub = (in_row*n_width+in_col)*n_depth+in_slice;
   if (ilayer == 0)
      value = f_inptr[isub];
```

```
    else
      value = actptr[isub];
    }
  s_cases[threadIdx.y][threadIdx.x] = value;
```

We compute the ordinal position of this term in the dot-product loop. For the weights, this was based on threadIdx.y, and for the activations it is based on threadIdx.x. The last block will be incomplete except in the unusual situation of the length of the dot product being an exact multiple of BLOCK_SIZE. If we are past the end of the dot product, the term is zero. And the last actual weight in the dot product is the bias, which by definition always has an activation of one.

We compute the position of this term in the prior layer's visual field exactly as we did for the weight. But because this is an actual prior-layer neuron, and not a filter weight that may be hanging over the edge into zero padding, we do not have to subtract the *base* position. Then just get the value, using the input image if this is the first hidden layer and using the prior hidden layer's activation if not.

All that's left to do is wait for the weight and activation loads to finish in all warps, sum the BLOCK_SIZE terms in this section of the dot product, wait for this computation to finish in all warps, apply the hyperbolic tangent activation function, and save the result.

```
    __syncthreads ();        // Wait for all shared memory loads to finish

    for (k=0; k<BLOCK_SIZE; k++)         // Sum these components
      sum += s_cases[threadIdx.y][k] * s_slices[k][threadIdx.x];

    __syncthreads ();        // Wait for the summation to finish in all warps

    } // For inner
  if (sum > MAX_EXP)      // Activation function
    sum = 1.0;
  else {
    sum = exp (2.0 * sum);
    sum = (sum - 1.0) / (sum + 1.0);
    }

  n_width = d_width[ilayer];
  n_depth = d_depth[ilayer];
```

```
    actptr = d_act[ilayer];              // Its activations are here
    ihid = (iheight * n_width + iwidth) * n_depth + idepth;    // Ordered (height, width, depth)
    actptr[icase*d_nhid[ilayer]+ihid] = sum;
}
```

There is one thing to note about storing the computed activation. Because depth changes fastest in the activation vector and idepth varies with threadIdx.x, at worst this store will be very well coalesced. If BLOCK_SIZE, n_depth, and d_nhid[ilayer] are all multiples of 16 (activations are double, not float), the stores will be perfectly coalesced. This, again, is strong motivation for the user to choose such values in the architecture.

Launch Code

At the start of this discussion, we saw a short code fragment illustrating how the shared-memory version of activation is performed in the host code. It's actually more complicated, largely because we cannot count on the dimensions of Figure 3-1 being an exact integer multiple of BLOCK_SIZE. We now discuss the launch code. It begins as shown here. In case we don't have enough slices or batch cases, use the non-shared-memory version that we saw on page 81.

```
int cuda_hidden_activation_LOCAL_CONV_shared (
    int local_vs_conv,       // Is this a LOCAL (vs CONV) layer?
    int istart,              // First case in this batch
    int istop,               // One past last case
    int nhid,                // Number of hidden neurons in this layer
    int n_slices,            // Depth of this layer
    int ilayer               // Layer to process
    )
{
    int nc, warpsize, threads_per_block;
    dim3 thread_launch, block_launch;
    cudaError_t error_id;

/*
    If possible (it normally would be), handle as much as possible with the more efficient
    shared-memory method.
    But if not, just use the non-shared method.
*/
```

```
nc = istop - istart;

if (n_slices < BLOCK_SIZE || nc < BLOCK_SIZE)
  return cuda_hidden_activation_LOCAL_CONV (
                        local_vs_conv, istart, istop, nhid, n_slices, ilayer);
```

The launch code for handling the complete set of blocks that fits within the entire set of slices and cases is simple.

```
thread_launch.x = BLOCK_SIZE;
thread_launch.y = BLOCK_SIZE;
thread_launch.z = 1;

block_launch.x = n_slices / BLOCK_SIZE;
block_launch.y = nc / BLOCK_SIZE;
block_launch.z = nhid / n_slices;      // Height times width; visual field size

device_hidden_activation_LOCAL_CONV_shared
        <<< block_launch, thread_launch >>> (local_vs_conv, istart, ilayer);

cudaDeviceSynchronize();
```

Now we deal with the slight complication of the slices and cases possibly exceeding a multiple of BLOCK_SIZE. This excess is illustrated in Figure 3-1. We use the non-shared-memory version presented on page 81 to clean up the extraneous slices and cases. First, we handle the entire right (slices) overhang, top to bottom.

```
if (n_slices % BLOCK_SIZE) {                      // Is there any overhang?
  threads_per_block = n_slices % BLOCK_SIZE;   // This much overhang
  block_launch.x = 1;
  block_launch.y = nhid / n_slices;   // Height times width; visual field size
  block_launch.z = nc;                            // All cases, top to bottom

  device_hidden_activation_LOCAL_CONV
      <<< block_launch, threads_per_block >>>
      (local_vs_conv, istart, 0,
      n_slices / BLOCK_SIZE * BLOCK_SIZE, n_slices % BLOCK_SIZE, ilayer);

  cudaDeviceSynchronize();
```

Then we clean up the bottom (cases) overhang. Because when we did the slices a moment ago and we went all the way to the bottom, we only do the rectangle directly below the blocks.

```
if (nc % BLOCK_SIZE) {                    // Is there any overhang?
   warpsize = deviceProp.warpSize;        // Threads per warp, likely 32 forever
   threads_per_block = (n_slices / BLOCK_SIZE * BLOCK_SIZE + warpsize - 1) /
                 warpsize * warpsize;     // Slices covered by blocks
   if (threads_per_block > 4 * warpsize)
      threads_per_block = 4 * warpsize;

   block_launch.x = (n_slices / BLOCK_SIZE * BLOCK_SIZE + threads_per_block - 1) /
                 threads_per_block;
   block_launch.y = nhid / n_slices;      // Height times width; visual field size
   block_launch.z = nc % BLOCK_SIZE;

   device_hidden_activation_LOCAL_CONV
        <<< block_launch, threads_per_block >>>
        (local_vs_conv, istart, nc / BLOCK_SIZE * BLOCK_SIZE, 0, n_slices /
        BLOCK_SIZE * BLOCK_SIZE, ilayer);

   cudaDeviceSynchronize();

   return 0;
}
```

Activating a Pooled Layer

Activating a pooled layer is much easier than what we've seen in the past two sections, for two reasons. First, there are no optimizable weights to load from global memory; the mapping function is fixed. Second, zero padding is not used, meaning that we don't have to deal with complex logic for handling edges. We begin with the launch code.

```
int cuda_hidden_activation_POOLED (
   int avg_vs_max,        // Is this a POOLAVG (vs POOLMAX) layer?
   int istart,            // First case in this batch
   int istop,             // One past last case
   int nhid,              // Number of hidden neurons in this layer
```

```
  int n_slices,              // Depth of this layer
  int ilayer                 // Layer to process
  )
{
  int warpsize, threads_per_block;
  dim3 block_launch;
  cudaError_t error_id;

  warpsize = deviceProp.warpSize;      // Threads per warp, likely 32 well into the future

  threads_per_block = (n_slices + warpsize - 1) / warpsize * warpsize;
  if (threads_per_block > 4 * warpsize)
    threads_per_block = 4 * warpsize;

  block_launch.x = (n_slices + threads_per_block - 1) / threads_per_block;
  block_launch.y = nhid / n_slices; // Height times width; visual field size
  block_launch.z = istop - istart;

  device_hidden_activation_POOLED <<< block_launch, threads_per_block >>>
              (avg_vs_max, istart, istop, ilayer);

  cudaDeviceSynchronize();

  return 0;
}
```

We see in the launch code that the thread determines the slice computed in the current layer. The position in the current layer's visual field is encoded into the y block coordinate as we've done before, and the case is in the block z coordinate. Here is the device code:

```
__global__ void device_hidden_activation_POOLED (
  int avg_vs_max,            // Is this a POOLAVG (vs POOLMAX) layer?
  int istart,                // First case in this batch
  int ilayer                 // Layer to process
  )
{
  int icase, iheight, iwidth, idepth, n_width, n_depth, ihid;
  int rstart, rstop, cstart, cstop, in_row, in_col, *poolmax_id_ptr;
  float *f_inptr;
  double x, *actptr, value;
```

```
idepth = blockIdx.x * blockDim.x + threadIdx.x;

if (idepth >= d_depth[ilayer])
    return;

n_width = d_width[ilayer];
n_depth = d_depth[ilayer];

iheight = blockIdx.y / n_width;      // Decode position in visual field
iwidth = blockIdx.y % n_width;
ihid = (iheight * n_width + iwidth) * n_depth + idepth; // Ordinal position in layer
```

We are about to compute the activation of the neuron at coordinates (iheight, iwidth, idepth) and ordinal position ihid in this layer. Note that it is critical that idepth be associated with the thread. This ensures that adjacent threads reference the same input, which allows efficient memory use. Why? When the thread advances, the position in the current layer's visual field does not change, and hence the rectangle referenced in the prior layer does not move. When an input for the first thread in a warp is loaded from global memory, this load is broadcast to the entire warp, saving all those other global loads.

```
icase = blockIdx.z;
```

We compute the position in the prior layer of the rectangle, which determines the activation of the neuron in the current layer. This is simple because we don't have to worry about edge effects from padding.

```
rstart = d_strideV[ilayer] * iheight;
rstop = rstart + d_PoolWidV[ilayer] - 1;
cstart = d_strideH[ilayer] * iwidth;
cstop = cstart + d_PoolWidH[ilayer] - 1;
```

As was the situation for earlier activation in the general case, we have to duplicate the same code for the first hidden layer (fed by the input) versus a subsequent hidden layer (fed by prior activations). This is because the input uses a float pointer, and activations use a double pointer. Deciding in the inner loop would be too slow.

```
if (ilayer == 0) {           // First hidden layer, so fed by input image
    f_inptr = d_predictors + (icase + istart) * d_n_pred;
```

```
    if (avg_vs_max) {
      value = 0.0;            // Will sum for average here
      for (in_row=rstart; in_row<=rstop; in_row++) { // Sum the rectangle
        for (in_col=cstart; in_col<=cstop; in_col++)
          value += f_inptr[(in_row*d_img_cols+in_col)*d_img_bands+idepth];
        } // For in_row
      value /= d_PoolWidV[ilayer] * d_PoolWidH[ilayer];
      }

    else {
      poolmax_id_ptr = &d_poolmax_id[ilayer][ihid] + icase * d_nhid[ilayer];
      value = -1.e60;          // Will keep track of max here
      for (in_row=rstart; in_row<=rstop; in_row++) {    // Check rectangle for max
        for (in_col=cstart; in_col<=cstop; in_col++) {
          x = f_inptr[(in_row*d_img_cols+in_col)*d_img_bands+idepth];
          if (x > value) {
            value = x;
            *poolmax_id_ptr = in_row * d_img_cols + in_col;   // Save id of max
            }
          } // For in_col
        } // For in_row
      } // POOLMAX
    } // If first hidden layer
```

As we did in the serial code on page 37, for max pooling we save the ID of the neuron in the prior layer, which was the rectangle max. This will prove handy when we backpropagate deltas from the pooling layer. Here is the rest of the device code, which essentially duplicates the previous code. In the last line, when we save the computed activation, note that ihid varies with idepth, which in turn varies with threadIdx.x. As a result, we are guaranteed at least very good coalescing, and sometimes perfect.

```
  else {
    actptr = d_act[ilayer-1] + icase * d_nhid[ilayer-1]; // Activation vector of prior layer
    n_width = d_width[ilayer-1];      // Size of prior layer
    n_depth = d_depth[ilayer-1];

    if (avg_vs_max) {
      value = 0.0;
```

```
      for (in_row=rstart; in_row<=rstop; in_row++) {
        for (in_col=cstart; in_col<=cstop; in_col++)
          value += actptr[(in_row*n_width+in_col)*n_depth+idepth];
        } // For in_row
      value /= d_PoolWidV[ilayer] * d_PoolWidH[ilayer];
      }

    else {
      poolmax_id_ptr = &d_poolmax_id[ilayer][ihid] + icase * d_nhid[ilayer];
      value = -1.e60;
      for (in_row=rstart; in_row<=rstop; in_row++) {
        for (in_col=cstart; in_col<=cstop; in_col++) {
          x = actptr[(in_row*n_width+in_col)*n_depth+idepth];
          if (x > value) {
            value = x;
            *poolmax_id_ptr = in_row * d_width[ilayer-1] + in_col; // Save id of max
            }
          } // For in_col
        } // For in_row
      } // POOLMAX
    }

  actptr = d_act[ilayer];
  actptr[icase*d_nhid[ilayer]+ihid] = value;
}
```

SoftMax and Log Likelihood by Reduction

The output activation routines compute only the logit of each output neuron. We must call a separate routine to do the SoftMax conversion. Then, we use a fancy reduction-based algorithm to compute the log likelihood function for the entire training set. SoftMax conversion is almost trivial, so we will gloss over it with just a token presentation of the code and a very few words of explanation. And log likelihood by reduction is covered in great depth in Volume 1 of this series. Because this topic is quite complex, there is no point in wasting paper by reproducing that long discussion. As with SoftMax, this section will do just a token presentation of the subject, trusting that confused readers will consult Volume 1 for clarification.

The launch code for SoftMax conversion is as follows, and the device code follows:

```
int cuda_softmax (
  int istart,    // First case in this batch
  int istop      // One past last case
  )
{
  int n, warpsize, blocks_per_grid, threads_per_block;
  cudaError_t error_id;

  warpsize = deviceProp.warpSize;    // Threads per warp, likely 32 well into the future

  n = istop - istart;    // Number of cases

  threads_per_block = (n + warpsize - 1) / warpsize * warpsize;
  if (threads_per_block > 4 * warpsize)
    threads_per_block = 4 * warpsize;

  blocks_per_grid = (n + threads_per_block - 1) / threads_per_block;

  device_softmax <<< blocks_per_grid, threads_per_block >>> (istart, istop);
  cudaDeviceSynchronize();
  return 0;
}
```

This is a simple one-dimensional launch. We pass the starting and stopping cases as parameters because output activations are stored for all training cases, not just those in the subset being processed. Thus, we need the starting case to properly offset the computation, and we need the stopping case so we know how many to do.

Note that the *nVidia* development system allows several types of exponentiation, which have trade-offs in speed and accuracy. However, speed is not a consideration here because this routine takes up an extremely small fraction of total computation time.

```
__global__ void device_softmax (
  int istart,     // First case in this batch
  int istop       // One past last case
  )
{
  int icase, iout;
  double *outptr, sum;
```

```
  icase = blockIdx.x * blockDim.x + threadIdx.x;

  if (icase >= istop - istart)
    return;

  outptr = d_output + (icase + istart) * d_n_classes;     // Output vector for this case
  sum = 0.0;

  for (iout=0; iout<d_n_classes; iout++) {
    if (outptr[iout] < MAX_EXP)                           // Do not allow disastrous overflow
      outptr[iout] = exp (outptr[iout]);
    else
      outptr[iout] = exp (MAX_EXP);
    sum += outptr[iout];
    }

  for (iout=0; iout<d_n_classes; iout++)
    outptr[iout] /= sum;
}
```

Here is the launch code for log likelihood computation. The number of threads, REDUC_THREADS, *must* be a power of two. The number of blocks given here, REDUC_BLOCKS, is a maximum. The actual number at runtime may be less. Note that reduc_fdata is a float array REDUC_BLOCKS long, allocated during initialization.

```
#define REDUC_THREADS 256
#define REDUC_BLOCKS 64

int cuda_ll (
  int n,         // Number of values; n_cases
  double *ll     // Computed log likelihood returned here
  )
{
  int i, blocks_per_grid;
  double sum;
  cudaError_t error_id;
```

```
blocks_per_grid = (n + REDUC_THREADS - 1) / REDUC_THREADS;
if (blocks_per_grid > REDUC_BLOCKS)
  blocks_per_grid = REDUC_BLOCKS;

device_ll <<< blocks_per_grid, REDUC_THREADS >>> ();
cudaDeviceSynchronize();

error_id = cudaMemcpy (reduc_fdata, h_ll_out, blocks_per_grid * sizeof(float),
                       cudaMemcpyDeviceToHost);

sum = 0.0;
for (i=0; i<blocks_per_grid; i++)
  sum += reduc_fdata[i];
*ll = sum;

return 0;
}
```

The device code will be completely cryptic to most readers who are not familiar with the technique of parallel reduction. I'll briefly discuss it here, but if this explanation is not enough, readers should see Volume 1 of this series for a long, detailed, step-by-step explanation.

Reduction happens in three distinct steps. In the first step, the threads cooperatively sum the individual case log likelihoods in big jumps spanning *threads per block* times *number of blocks*, as the total number of cases will usually exceed this product. The partial sum for each thread is stored in fast shared memory. The second step crunches these partial sums pairwise, halving their number with each pass through the loop. The third step is performed in the launch code shown earlier; it does the final summation.

```
__global__ void device_ll ()
{
  __shared__ double partial_ll[REDUC_THREADS];
  int i, n, n_classes, index;
  double sum_ll;

  index = threadIdx.x;
  n = d_ncases;
  n_classes = d_n_classes;
```

```
   sum_ll = 0.0;
   for (i=blockIdx.x*blockDim.x+index; i<n; i+=blockDim.x*gridDim.x)
      sum_ll -= log (d_output[i*n_classes+d_class[i]] + 1.e-30);

   partial_ll[index] = sum_ll;
   __syncthreads();

   for (i=blockDim.x>>1; i; i>>=1) {
      if (index < i)
         partial_ll[index] += partial_ll[index+i];
      __syncthreads();
      }

   if (index == 0)
      d_ll_out[blockIdx.x] = partial_ll[0];
}
```

Computing Delta for the Output Layer

The routine for computing the output delta vector and placing it in this_delta is almost too trivial to list in the book, but here it is for reference. The launch code is first, followed by the device code.

```
int cuda_output_delta (
   int istart,    // First case in this batch
   int istop,     // One past last case
   int ntarg      // Number of targets (outputs, classes)
   )
{
   int warpsize, threads_per_block;
   dim3 block_launch;
   cudaError_t error_id;

   warpsize = deviceProp.warpSize;     // Threads per warp, likely 32 well into the future

   threads_per_block = (ntarg + warpsize - 1) / warpsize * warpsize;
   if (threads_per_block > 4 * warpsize)
      threads_per_block = 4 * warpsize;
```

```
block_launch.x = (ntarg + threads_per_block - 1) / threads_per_block;
block_launch.y = istop - istart;
block_launch.z = 1;

device_output_delta <<< block_launch, threads_per_block >>> (istart);

cudaDeviceSynchronize();
return 0;
}
```

In the previous code, you see that the threads are used for the output neurons. In any other case this would be silly because in most applications there are not enough classes to fill even one warp! But it's the simplest approach, and efficiency is unimportant because this routine takes up an almost unmeasurably small fraction of the total run time.

The device code is nothing more than a straightforward implementation of Equation 1-12. The following things should be noted:

- During initialization, the d_class vector was computed. This is the integer (zero origin) class ID of every case in the training set. This code does not appear in the book, but it can be found in the file MOD_CUDA.cu.

- The d_output vector contains outputs for every case in the training set. Thus, its index must be offset by istart, the first case in the batch being processed.

- Like most other device-memory storage, d_this_delta contains delta for only those cases in the batch being processed. Thus, its index is *not* offset by istart.

- Both d_output and d_this_delta are ordered with the output neuron (which, for any fully connected layer, is the depth) changing fastest. Therefore, memory accesses for both are very well coalesced.

```
__global__ void device_output_delta (
  int istart      // First case in this batch
  )
{
  int icase, iout;
  double target;
```

```
iout = blockIdx.x * blockDim.x + threadIdx.x;

if (iout >= d_n_classes)
    return;

icase = blockIdx.y;
target = (iout == d_class[istart+icase]) ? 1.0 : 0.0;

d_this_delta[icase*d_n_classes+iout] =
                        target - d_output [ (istart + icase) * d_n_classes + iout ];
}
```

Backpropagating from a Fully Connected Layer

This section presents code for backpropagating delta from a fully connected layer to a prior layer of any type. It has a simple two-dimensional launch. Each case has its own block or set of blocks. The thread in a block is associated with the hidden neuron in the *receiving* layer, the layer *prior to* the fully connected layer whose delta already exists. In the code to follow, layer ilayer is receiving the backpropagated delta, and ilayer+1 is the fully connected layer. Here is the launch code:

```
int cuda_backprop_delta_FC (
    int nc,              // Number of cases in batch
    int ilayer,          // Hidden layer being processed
    int nhid_this        // Number of hidden neurons in this layer
    )
{
    int warpsize, threads_per_block;
    dim3 block_launch;
    cudaError_t error_id;

    warpsize = deviceProp.warpSize;     // Threads per warp, likely 32 well into the future

    threads_per_block = (nhid_this + warpsize - 1) / warpsize * warpsize;
    if (threads_per_block > 4 * warpsize)
        threads_per_block = 4 * warpsize;

    block_launch.x = (nhid_this + threads_per_block - 1) / threads_per_block;
    block_launch.y = nc;
    block_launch.z = 1;
```

```
  device_backprop_delta_FC <<< block_launch, threads_per_block >>> (ilayer);

  cudaDeviceSynchronize ();
  return 0;
}
```

Here is the device code. Comments will be interspersed.

```
__global__ void device_backprop_delta_FC (
  int ilayer     // Feed is from ilayer to ilayer+1, so ilayer+1 is fully connected
  )
{
  int j, icase, ihid, nhid, n_next;
  float *next_weights;
  double *delta_ptr, *prior_delta_ptr, this_act, delta;

  ihid = blockIdx.x * blockDim.x + threadIdx.x;
  nhid = d_nhid[ilayer];      // Neurons in this hidden layer

  if (ihid >= nhid)
    return;

  icase = blockIdx.y;
```

We now get the number of neurons in the next layer, and a pointer to the weight vector connecting the current layer to the next layer. Recall that to achieve perfect coalescing for the often-used weights, they are zero padded to multiples of 128 bytes. This is why we multiply by d_nhid_cols and d_n_classes_cols, which are the padded sizes. This topic is discussed on page 72. Unfortunately, this destroys coalescing in this particular routine. Fortunately, this routine generally requires only a tiny fraction of total application time, so speed is not important. Moreover, the heavy double-precision math does an excellent job of hiding access times. So it's no problem at all.

```
  if (ilayer == d_n_layers-1) {       // Next layer is the output layer?
    n_next = d_n_classes;
    next_weights = d_weights[ilayer+1] + ihid * d_n_classes_cols;
    }
```

```
else {                              // Next layer is another hidden layer
  n_next = d_nhid[ilayer+1];
  next_weights = d_weights[ilayer+1] + ihid * d_nhid_cols[ilayer+1];
  }
```

At this time, d_this_delta is delta for the next layer, already computed. We now compute d_prior_delta. These arrays are not zero padded because their accesses are well coalesced and not very speed critical.

```
delta_ptr = d_this_delta + icase * n_next;      // This already exists
prior_delta_ptr = d_prior_delta + icase * nhid;   // This is being computed now
```

The next few lines are a direct implementation of Equation 1-19. The loop is the summation part of this equation. Then, for layers that have a nonlinear activation function, we complete the equation by multiplying by the derivative of the activation function. This derivative was given by Equation 1-15.

```
delta = 0.0;
for (j=0; j<n_next; j++)
  delta += delta_ptr[j] * next_weights[j];

if (d_layer_type[ilayer] == TYPE_FC ||
    d_layer_type[ilayer] == TYPE_LOCAL ||
    d_layer_type[ilayer] == TYPE_CONV) {
  this_act = d_act[ilayer][icase*nhid+ihid];
  delta *= 1.0 - this_act * this_act;      // Derivative; Equation 1-15 on Page 20
  }

prior_delta_ptr[ihid] = delta;          // Save it for doing the next layer back
}
```

Backpropagating from Convolutional and Local Layers

When we presented code for backpropagation from convolutional and locally connected layers back on page 53, we reversed the summation of Equation 1-19, as this was the most efficient way of handling the operation in serial code. But in parallel CUDA code, it is more efficient to perform the summation directly because each thread handles a single neuron in the current layer.

Here is the simple launch code. Recall that we are computing delta for layer ilayer, using existing deltas from layer ilayer+1, which is convolutional or locally connected.

```
int cuda_backprop_delta_nonpooled (
   int nc,          // Number of cases in batch
   int ilayer,      // Hidden layer being processed, based on ilayer+1
   int nhid_this    // Number of hidden neurons in this layer
   )
{
   int warpsize, threads_per_block;
   dim3 block_launch;
   cudaError_t error_id;

   warpsize = deviceProp.warpSize;      // Threads per warp, likely 32 well into the future

   threads_per_block = (nhid_this + warpsize - 1) / warpsize * warpsize;
   if (threads_per_block > 4 * warpsize)
      threads_per_block = 4 * warpsize;

   block_launch.x = (nhid_this + threads_per_block - 1) / threads_per_block;
   block_launch.y = nc;
   block_launch.z = 1;

   device_backprop_delta_nonpooled <<< block_launch, threads_per_block >>> (ilayer);

   cudaDeviceSynchronize();

   return 0;
}
```

The device code is a straightforward implementation of Equation 1-19. But it does have some complexity that is due to reversing the mapping from a layer to the next. It's easy to take a neuron in a given layer and determine the neurons in the *prior* layer that are in the activation rectangle; we've done it several times already. But it's not so easy to take a neuron in a given layer and figure out which neurons in the *next* layer are fed by it. Most of the device code for this routine is devoted to this task. Here is the calling parameter list and the beginning of the routine:

```
__global__ void device_backprop_delta_nonpooled (
   int ilayer // Feed is from ilayer to ilayer+1, so ilayer+1 is LOCAL or CONV
   )
```

```
{
    int k, icase, ihid, next_row, next_col, next_slice, this_row, this_col, this_slice;
    int nH, k_next, wt_cols, rstart, cstart, prod, ltype;
    int strideH, strideV, padH, padV, height, width, depth;
    int next_rstart, next_rstop, next_cstart, next_cstop;
    float *weights, *wtptr;
    double *this_delta_ptr, *prior_delta_ptr, this_act, sum;

    ihid = blockIdx.x * blockDim.x + threadIdx.x;

    if (ihid >= d_nhid[ilayer])
        return;
```

This first block of code gets the (row, column, slice) coordinates of neuron ihid. This is the neuron whose delta we are about to compute. Then we get the case and compute the horizontal width of the filter that connects this layer to the next.

```
prod = d_width[ilayer] * d_depth[ilayer];
this_row = ihid / prod;
k = ihid - this_row * prod;
this_col = k / d_depth[ilayer];
this_slice = k % d_depth[ilayer];

icase = blockIdx.y;

nH = 2 * d_HalfWidH[ilayer+1] + 1; // Horizontal filter size
```

We now get pointers to the next layer's delta, which is known, and this layer's delta, which we will compute here. It's efficient to gather into registers architectural details that will be referenced often later.

```
this_delta_ptr = d_this_delta + icase * d_nhid[ilayer+1];
prior_delta_ptr = d_prior_delta + icase * d_nhid[ilayer];

ltype = d_layer_type[ilayer+1];
strideV = d_strideV[ilayer+1];
strideH = d_strideH[ilayer+1];
padV = d_padV[ilayer+1];
padH = d_padH[ilayer+1];
height = d_height[ilayer+1];
width = d_width[ilayer+1];
depth = d_depth[ilayer+1];
```

The next few lines of code are the crux of reversing the mapping direction. Please understand the two comments that precede the code. We can do this in integer arithmetic. If necessary, review the section that starts on page 31 to understand that when we look back to the prior layer, the activation rectangle begins at the current coordinate, times the stride, minus the pad, and ends at twice the half-width later (inclusive). We (carefully!) reverse this direction, especially noting that if the division for the start was inexact, we must bypass the fractional part.

```
// this >= next * stride - pad IMPLIES next <= (this + pad) / stride
// this <= next * stride - pad + 2 * hw IMPLIES next >= (this + pad - 2 * hw) / stride

next_rstop = this_row + padV;
k = next_rstart = next_rstop - 2 * d_HalfWidV[ilayer+1];
next_rstop /= strideV;
next_rstart /= strideV;

if (k >= 0 && k % strideV)          // If the division above was inexact
   ++next_rstart;                   // We must move past fractional part

if (next_rstop >= height)           // Stay inside the visual field
   next_rstop = height - 1;
if (next_rstart < 0)
   next_rstart = 0;
next_cstop = this_col + padH;
k = next_cstart = next_cstop - 2 * d_HalfWidH[ilayer+1];
next_cstop /= strideH;
next_cstart /= strideH;
if (k >= 0 && k % strideH)
   ++next_cstart;
if (next_cstop >= width)
   next_cstop = width - 1;
if (next_cstart < 0)
   next_cstart = 0;
```

Get a pointer to the weights that connect this layer to the next layer. We need to know the length of these padded weight vectors. A convolutional layer has the same weight set for every neuron in the visual field of a given slice, so weights change only with the slice. But a locally connected layer has a different weight set for every neuron in the layer. Then we zero the sum that will cumulate delta.

```
weights = d_weights[ilayer+1];
if (ltype == TYPE_CONV)
  wt_cols = d_depth_cols[ilayer+1];
else
  wt_cols = d_nhid_cols[ilayer+1];

sum = 0.0;
```

Thanks to reversing the order of rectangle definition, which we did earlier, we know the exact limits of the rectangle in the next layer to which the current neuron connects. Thus, we can limit our summation to this rectangle. We do need the starting coordinates of the rectangle in the *current* layer so that we can compute the position of the *current* neuron in the filter. We've seen this simple formula many times before!

```
for (next_row=next_rstart; next_row<=next_rstop; next_row++) {
  for (next_col=next_cstart; next_col<=next_cstop; next_col++) {

    // Center of first filter is at HalfWidth-Pad; filter begins at -Pad.
    rstart = strideV * next_row - padV;
    cstart = strideH * next_col - padH;
    // This is what we would be testing if we didn't compute the exact limits above
    // rstop = rstart + 2 * d_HalfWidV[ilayer+1];
    // cstop = cstart + 2 * d_HalfWidH[ilayer+1];
    // if (this_row>=rstart && this_row<=rstop && this_col>=cstart && this_col<=cstop){

    for (next_slice=0; next_slice<depth; next_slice++) {
```

As a point of interest, those last few commented-out lines show what we would be doing if we had not reversed the rectangle direction to get exact limits. It would be significantly more work.

Here is the last bit of cryptic computation. We compute k_next as the ordinal position of the neuron in the next layer that we are handling in this triply-nested loop. This identifies the starting weight for locally connected layers. But because convolutional layers share the same weight set for all neurons in a given slice, its weight set is determined by the slice alone. Note that efficiency could be slightly improved, at the cost of slightly less clarity, if we move some aspects of these computations earlier in the nested loops to avoid repetition. Confident readers may want to do so.

We compute k as the location in the filter of the weight that connects neuron ihid in the current layer to neuron k_next in the next layer. The product of this weight times the delta of that next-layer neuron is a single term in the summation of Equation 1-19.

```
        k_next = (next_row * width + next_col) * depth + next_slice;
        if (ltype == TYPE_CONV)
          wtptr = weights + next_slice;
        else
          wtptr = weights + k_next;
        k = ((this_row - rstart) * nH + this_col - cstart) * d_depth[ilayer] + this_slice;
        sum += this_delta_ptr[k_next] * wtptr[k*wt_cols];
        } // For next_col
      } // For next_row
    } // For next_slice
```

We are almost finished. The last step is to complete that equation by multiplying the sum by the derivative of the activation of the current neuron. Note that when we save the computed delta, the subscript is based on threadIdx.x, so the save is well coalesced.

```
  if (d_layer_type[ilayer] == TYPE_FC ||
      d_layer_type[ilayer] == TYPE_LOCAL ||
      d_layer_type[ilayer] == TYPE_CONV) {
    this_act = d_act[ilayer][icase*d_nhid[ilayer]+ihid];
    sum *= 1.0 - this_act * this_act;    // Derivative
    }
  prior_delta_ptr[ihid] = sum;
}
```

Astute readers will observe that accesses to the weight vector are very poorly coalesced. This is the price paid for perfect coalescing when the activation is computed. It's a great trade-off because in virtually all applications, the time spent computing activations is tremendously greater than the time spent backpropagating delta, often several orders of magnitude greater. So this inefficient weight access here is of no practical consequence.

Backpropagating from a Pooling Layer

The algorithm for backpropagating from a pooling layer is similar to that shown in the prior section. Thus, we will gloss over most explanations and focus on the few differences. Here is the simple launch code, which is virtually identical to that of the prior section:

```
int cuda_backprop_delta_pooled (
   int nc,            // Number of cases in batch
   int ilayer,        // Hidden layer being processed
   int nhid_this      // Number of hidden neurons in this layer
   )
{
   int warpsize, threads_per_block;
   dim3 block_launch;
   cudaError_t error_id;

   warpsize = deviceProp.warpSize;     // Threads per warp, likely 32 well into the future

   threads_per_block = (nhid_this + warpsize - 1) / warpsize * warpsize;
   if (threads_per_block > 4 * warpsize)
      threads_per_block = 4 * warpsize;

   block_launch.x = (nhid_this + threads_per_block - 1) / threads_per_block;
   block_launch.y = nc;
   block_launch.z = 1;

   device_backprop_delta_pooled <<< block_launch, threads_per_block >>> (ilayer);
   cudaDeviceSynchronize();

   return 0;
}
```

The device code is so similar at first to that in the prior section that we will list everything up to the point of difference here, without explanation. See the prior section as needed.

```
__global__ void device_backprop_delta_pooled (
   int ilayer // Feed is from ilayer to ilayer+1, so ilayer+1 is POOLAVG or POOLMAX
   )
{
```

```
int k, icase, ihid, next_row, next_col, this_row, this_col, this_slice;
int k_next, prod, this_cols, *poolmax_id_ptr;
int next_rstart, next_rstop, next_cstart, next_cstop;
double *this_delta_ptr, *prior_delta_ptr, sum, this_act;

ihid = blockIdx.x * blockDim.x + threadIdx.x;

if (ihid >= d_nhid[ilayer])
   return;

prod = d_width[ilayer] * d_depth[ilayer]; // Get the 3D coordinates of this neuron
this_row = ihid / prod;
k = ihid - this_row * prod;
this_col = k / d_depth[ilayer];
this_slice = k % d_depth[ilayer];

icase = blockIdx.y;

this_delta_ptr = d_this_delta + icase * d_nhid[ilayer+1];      // Coming from next layer
prior_delta_ptr = d_prior_delta + icase * d_nhid[ilayer];      // Will compute this

// this >= next * stride IMPLIES next <= this / stride
// this <= next * stride + pw - 1 IMPLIES next >= (this - pw + 1) / stride
// We can safely do this in integer arithmetic

next_rstop = this_row;
k = next_rstart = next_rstop - d_PoolWidV[ilayer+1] + 1;
next_rstop /= d_strideV[ilayer+1];
next_rstart /= d_strideV[ilayer+1];
if (k >= 0 && k % d_strideV[ilayer+1])
   ++next_rstart;
if (next_rstop >= d_height[ilayer+1])
   next_rstop = d_height[ilayer+1] - 1;
if (next_rstart < 0)
   next_rstart = 0;

next_cstop = this_col;
k = next_cstart = next_cstop - d_PoolWidH[ilayer+1] + 1;
next_cstop /= d_strideH[ilayer+1];
next_cstart /= d_strideH[ilayer+1];
```

```
if (k >= 0 && k % d_strideH[ilayer+1])
   ++next_cstart;

if (next_cstop >= d_width[ilayer+1])
   next_cstop = d_width[ilayer+1] - 1;
if (next_cstart < 0)
   next_cstart = 0;

sum = 0.0;
```

Here is where this routine differs from the prior routine. We handle average pooling first. We don't have to worry about weights because the weights are fixed, not trainable. If this is kept in mind, we see that the algorithm is practically identical to that seen in the prior section. Just remember that a pooling layer maps slice by slice from the prior layer.

```
if (d_layer_type[ilayer+1] == TYPE_POOLAVG) {
   for (next_row=next_rstart; next_row<=next_rstop; next_row++) {
     for (next_col=next_cstart; next_col<=next_cstop; next_col++) {
       k_next = (next_row*d_width[ilayer+1] + next_col)*d_depth[ilayer+1] + this_slice;
       sum += this_delta_ptr[k_next];
       } // For next_col
     } // For next_row
   sum /= d_PoolWidH[ilayer+1] * d_PoolWidV[ilayer+1];
   } // POOLAVG
```

The other possibility is that this is max pooling. This is slightly more complex because exactly one of the "weights," that which corresponds to the maximum activation in the prior-layer rectangle, is 1.0, and all other weights are zero. Recall that when we computed the activations (page 101) we saved in d_poolmax_id the position in the visual field of the winning prior-layer neuron. Now we see why this was a good move.

We get a pointer to this saved information. As we did in average pooling, we loop through the visual field of the next layer. For each neuron in the set of possible connections, we check to see whether the neuron in the current layer is the winner in the competition that determined the activation of the neuron in the next layer. If so, the "weight" is 1.0. Otherwise, the weight is zero.

```
  else if (d_layer_type[ilayer+1] == TYPE_POOLMAX) {
    poolmax_id_ptr = d_poolmax_id[ilayer+1] + icase * d_nhid[ilayer+1];
    this_cols = d_width[ilayer];
    for (next_row=next_rstart; next_row<=next_rstop; next_row++) {
      for (next_col=next_cstart; next_col<=next_cstop; next_col++) {
        k_next = (next_row*d_width[ilayer+1] + next_col)*d_depth[ilayer+1] + this_slice;
        // Was the current-layer neuron the winner in the MAX competition
        // for the next-layer competition?
        if (this_row == poolmax_id_ptr[k_next] / this_cols &&
          this_col == poolmax_id_ptr[k_next] % this_cols)
          sum += this_delta_ptr[k_next]; // Weight is 1
        } // For next_col
      } // For next_row
    } // POOLMAX
```

Finally, we multiply by the derivative of the current layer's activation function and save the result.

```
  if (d_layer_type[ilayer] == TYPE_FC || d_layer_type[ilayer] == TYPE_LOCAL ||
d_layer_type[ilayer] == TYPE_CONV) {
    this_act = d_act[ilayer][icase*d_nhid[ilayer]+ihid];
    sum *= 1.0 - this_act * this_act;        // Derivative
    }
  prior_delta_ptr[ihid] = sum;               // Save it for doing the next layer back
}
```

Gradient of a Fully Connected Layer

This and the next few sections deal with computing the gradient. We will hold off on presenting the launch code until all layer types are covered. This is because we use a single gradient launch routine that selects the correct device code for each layer type.

All of the device routines implement the simple Equation 1-18, which just multiplies a neuron's delta by the activation of a prior- layer neuron to get the partial derivative of the performance criterion with respect to the connecting weight. We begin with the routine for a fully connected layer, as it is the easiest to understand.

```
__global__ void device_hidden_gradient_FC (
  int istart,    // Index of first case in this batch
  int nc,        // Number of cases in batch
  int ilayer     // Hidden layer being processed
  )
{
  int iin, ihid, nin, ninp1;
  float *gptr;
  double input;

  iin = blockIdx.x * blockDim.x + threadIdx.x;

  if (ilayer == 0)
    nin = d_n_pred;        // Number of inputs to each neuron in this layer
  else
    nin = d_nhid[ilayer-1];

  // icase = blockIdx.z;    // Used directly below

  if (iin > nin)
    return;
  else if (iin == nin)     // This is the bias term, which by definition is 1.0
    input = 1.0;
  else if (ilayer)         // The prior layer is a hidden layer, so get its activations
    input = d_act[ilayer-1][blockIdx.z*nin+iin];
  else                     // This is the first hidden layer, so its input is the input image
    input = d_predictors[(istart+blockIdx.z)*nin+iin];
  ihid = blockIdx.y;       // Ordinal number of this hidden neuron
  ninp1 = nin + 1;         // We mustn't forget the bias, so nin+1

  gptr = d_grad[ilayer] + blockIdx.z * d_n_weights; // Gradient of hidden layer for case
  gptr[ihid*ninp1+iin] = d_this_delta[blockIdx.z*d_nhid[ilayer]+ihid] * input;
}
```

It's worth noting that there are four global memory accesses.

- When we set input equal to a prior-layer activation, the memory offset is tied to threadIdx.x, so the read is very well coalesced.

- When we set input equal to an element of the input image, the memory offset is tied to threadIdx.x, so the read is very well coalesced.

- When we fetch this neuron's delta, the memory address is independent of the thread, so this single read value is broadcast to the entire warp, which is extremely efficient.

- When we store the computed value to the gradient vector, the memory offset for the store is tied to threadIdx.x, so the write is very well coalesced.

Gradient of a Locally Connected or Convolutional Layer

This routine conceptually does the same thing as the routine in the prior section. But the big difference is that most connecting weights are zero. Thus, it is incumbent on us to make sure to process the activation-times-delta products as efficiently as possible. This is especially true in that for most architectures, this routine is the dominant eater of compute time. Efficiency is of the utmost importance, especially in regard to global memory reads, which are prolific.

Here is the beginning of the device code. The calling parameters should all be self-explanatory, with one possible exception. This routine allows processing slices of the current layer in subsets; it does not demand that every neuron be processed at once. We will see later that it is sometimes necessary to break up computation into multiple launches, each launch processing one or more slices. The depth_offset parameter tells us where to begin processing (0 is the first slice), and n_depths tells us how many slices to process in this launch.

```
__global__ void device_hidden_gradient_LOCAL_CONV (
  int local_vs_conv,        // Is this a LOCAL (vs CONV) layer?
  int nfilt,                // Filter size, (2*hwV+1) * (2*hwH+1) * depth of input
                            // This does not include the +1 for the bias term
  int istart,               // Index of first case in this batch
  int depth_offset,         // Start processing layers at this depth
  int n_depths,             // Number of slices to be processed
  int ilayer                // Hidden layer being processed
  )
```

```
{
  int k, iin, ifilt, ihid_offset, ihid_actual, prod;
  int in_row, in_col, in_slice, in_rows, in_cols, in_slices;
  int this_row, this_col, ifiltV, ifiltH;
  float *gptr;
  double input, delta;

  ifilt = blockIdx.x * blockDim.x + threadIdx.x;  // <= filter size
  if (ifilt > nfilt)
    return;
```

We see in the previous code that threads correspond to weights in the prior-layer rectangle that, when dotted with the corresponding prior-layer activations, form the activation of the current neuron. If *hwV* and *hwH* are the filter half-widths, there are a total of $(2*hwV+1) * (2*hwH+1) * depth\ of\ prior\ layer$ such weights, plus one more weight for the bias term, feeding each neuron in the current layer. The launcher supplies this product, not including the +1 for the bias, in the nfilt parameter. Our first act is to get the dimensions of the volume feeding this layer.

```
if (ilayer == 0) {
  in_rows = d_img_rows;
  in_cols = d_img_cols;
  in_slices = d_img_bands;
  }
else {
  in_rows = d_height[ilayer-1];
  in_cols = d_width[ilayer-1];
  in_slices = d_depth[ilayer-1];
  }
```

The next few lines of code are a bit tricky. Recall that we may be starting gradient computation at some slice past the first. We get the offset *from* the first neuron in the first slice being processed *to* the neuron being processed in blockIdx.y. As we'll see in the launch code later, the maximum value of this quantity is guaranteed to be a multiple of the visual field size of the current layer, minus one. Thus, a launch will always process exactly n_depths times the visual field size neurons. No launch will ever process just part of the visual field.

```
ihid_offset = blockIdx.y;                          // Offset into this launch set
prod = d_width[ilayer] * d_height[ilayer];         // Size of visual field, a slice
k = ihid_offset % n_depths + depth_offset;         // Actual starting slice
ihid_actual = ihid_offset / n_depths * d_depth[ilayer] + k;
```

The code shown previously is necessary because we will be working with two different versions of *ihid*, the neuron in the current layer. We have ihid_offset, the offset into the subset of slices being processed in this launch, and we also have ihid_actual, the ordinal position in the entire layer. These four lines compute the actual starting slice, k, as the remainder from dividing the offset by the number of depths in this launch and then adding the offset to the first slice. Remember that neurons are ordered with depth changing fastest. Then we divide the offset by the number of depths to get the visual field position, multiply by the layer's depth to get the start of slices in this visual field position, and add the actual starting slice.

If this is not clear, imagine a chessboard with checkers stacked up in equal numbers on every square. You have a sheet of paper lying partway up the set of stacks. The board is the visual field, and the stacks of checkers are the slices. The sheet of paper marks the start of the set of slices being processed. Counting starts at the bottom layer at the top-left corner. It goes up the first stack, then moves on to the bottom of the next stack to the right, and so forth. Now work through the code with this image in mind.

Before continuing, we have to take a brief break to discuss the difference between gradient computation for convolutional layers versus locally connected layers. The former uses the same filter weight set for all neurons in the visual field of a given slice, while the latter uses a different weight set for every neuron. That latter situation is just a specialized version of a fully connected layer most of whose connection weights are zero, and hence computation is similar to what we've already seen. But the former situation is unusual in that perturbing a single weight will impact activations all across the visual field. How do we handle this complexification?

The good news is that the effects of minuscule perturbation are linear, so to compute the partial derivative with respect to a given weight, we simply compute the weight's partial derivative for every individual neuron in the visual field, exactly as if this were a locally connected layer, and add them.

The bad news is that the vector we use to store the gradient has slots for only the common set of weights. If we are going to use the same algorithm for both layer types (and this is the most efficient way to do it), then we need to have a work area for temporarily holding the individual gradients across the visual field. We'll compute them, store them in this work area, and then invoke a separate kernel to sum them. Allocation of this work area will be discussed later in this chapter. For now, assume that it exists. It is called d_convgrad_work, and its length is d_max_convgrad_each per case.

We can continue exploring this device routine now. If this thread is handling the bias term, things are simple. Recall that blockIdx.z is the case in this batch, d_n_prior_weights[ilayer] is the number of weights, and the bias term is the last entry in the weight vector. If this is a locally connected layer, we store the derivative (delta, because the activation of a bias term is 1) directly into the gradient vector. But if this is a convolutional layer, we store delta in the work area just discussed. For perfect coalescing, this work area is padded to a multiple of 128 bytes, and this padded length is d_convgrad_cols[ilayer].

```
if (ifilt == nfilt) { // Bias term
  delta = d_this_delta[blockIdx.z*d_nhid[ilayer]+ihid_actual];
  if (local_vs_conv) {
    gptr = d_grad[ilayer] + blockIdx.z * d_n_weights;
    gptr[ihid_actual*d_n_prior_weights[ilayer]+d_n_prior_weights[ilayer]-1] = delta;
    }
  else {
    gptr = d_convgrad_work + blockIdx.z * d_max_convgrad_each;
    gptr[ihid_offset*d_convgrad_cols[ilayer]+d_n_prior_weights[ilayer]-1] = delta;
    }
  return;
  }
```

If we get here, this is not the bias term. Get the location of this kernel within the filter. The thread defines ifilt, the ordinal number of the filter weight. Remember that the order of weight storage for the filter is (height, width, slice).

```
prod = (2 * d_HalfWidH[ilayer] + 1) * in_slices;        // This many elements per row
ifiltV = ifilt / prod;                   // Vertical position in filter
k = ifilt - ifiltV * prod;
ifiltH = k / in_slices;                  // Horizontal position in filter
in_slice = k % in_slices;                // Input slice to which this filter weight applies
```

Get the location of this neuron within the volume of the current layer.

```
prod = d_width[ilayer] * d_depth[ilayer];        // Size of current layer's visual field
this_row = ihid_actual / prod;                   // Row of current neuron
k = ihid_actual - this_row * prod;
this_col = k / d_depth[ilayer];                  // Column of current neuron
// this_slice = k % d_depth[ilayer];             // Not needed; here for clarity only
```

Now that we know the neuron in the current layer, and hence the corresponding rectangle in the prior (input) layer, we can get the location of this filter element within the input volume. Because of padding, it may be outside an edge, in which case there is nothing to do.

We have seen the basic math for locating the prior-layer rectangle several times before, but here it is once again in case you've forgotten:

- The filter center is at Stride * CurrentPos + HalfWidth - Pad.

- The upper-left corner is at Stride * CurrentPos - Pad.

```
in_row = d_strideV[ilayer] * this_row - d_padV[ilayer] + ifiltV;
if (in_row < 0 || in_row >= in_rows)             // Outside top or bottom edge
   return;

in_col = d_strideH[ilayer] * this_col - d_padH[ilayer] + ifiltH;
if (in_col < 0 || in_col >= in_cols)             // Outside left or right edge
   return;
```

We get a pointer to the place where we will put the computed derivative, exactly as we did for the bias term earlier. Also, we fetch delta from global memory. Note that the memory address of delta is independent of the thread, so this single value is efficiently broadcast to the entire warp with a single load.

```
if (local_vs_conv)
   gptr = d_grad[ilayer] + blockIdx.z * d_n_weights;
else
   gptr = d_convgrad_work + blockIdx.z * d_max_convgrad_each;

delta = d_this_delta[blockIdx.z*d_nhid[ilayer]+ihid_actual];
```

We've got delta, and we know where to put the derivative. Now we fetch the input corresponding to this filter weight. Adjacent threads have adjacent memory accesses, though not zero padded for alignment. But zero padding would do no good here because in the most general case warps will only by chance start properly aligned. So, in the worst case, coalescing will be very good. And if in_slices and the prior-layer size are both multiples of 16 (activities are double, not float), then coalescing will be perfect.

```
iin = (in_row * in_cols + in_col) * in_slices + in_slice;
if (ilayer)
   input = d_act[ilayer-1][blockIdx.z*d_nhid[ilayer-1]+iin];
else
   input = d_predictors[(istart+blockIdx.z)*d_n_pred+iin];
```

The last step is to store the computed gradient value. Adjacent threads access adjacent memory, so at worst, coalescing is very good. There is no zero padding of the gradient vector for alignment. Zero padding would help for locally connected layers, because ifilt starts at zero. But that would complicate the code a lot, and this is a small fraction of instructions. Also, the kernel is generally limited by the math pipeline. And of course if n_prior_weights is a multiple of 32, all is good! Finally, d_convgrad_work is padded properly, so for convolutional layers (which is mostly what we use!), coalescing is perfect.

```
if (local_vs_conv)
   gptr[ihid_actual*d_n_prior_weights[ilayer]+ifilt] = input * delta;
else
   gptr[ihid_offset*d_convgrad_cols[ilayer]+ifilt] = input * delta;
}
```

Flattening the Convolutional Gradient

We saw that for a convolutional layer, we store the gradient term of each individual neuron of a slice's visual field in a work area. Thus, we must sum them to get the gradient for the common filter weight set. Each slice has its own set of filter weights, so this summation is done separately for each slice in the current layer. Here is the beginning of the device routine for doing this. Just as was the case for computing the gradient, we allow here for the launch, processing just a subset of all slices in the current layer. Thus, islice_start is the index of the first slice to be processed, and max_depth is the number of slices to process in this launch.

```
__global__ void device_flatten_gradient (
  int islice_start,              // Index of first slice in this batch
  int max_depth,                 // Max slices in launch, <= slices reserved in convgrad_work
  int ilayer                     // Hidden layer being processed
  )
{
  int k, islice, icase, iprior, irow, icol;
  double sum;
  float *workptr, *gradptr;

  iprior = blockIdx.x * blockDim.x + threadIdx.x;
  if (iprior >= d_n_prior_weights[ilayer])
    return;

  islice = blockIdx.y;
  icase = blockIdx.z;
```

We see in the previous code that the thread determines the location in the filter rectangle that this thread will handle. The current-layer slice and the case come from the block. Get pointers to the gradient vector that will be computed and the work area that is to be flattened by summation. Initialize the sum for this thread to zero.

```
  gradptr = d_grad[ilayer] + icase * d_n_weights;
  workptr = d_convgrad_work + icase * d_max_convgrad_each;

  sum = 0.0;
```

The final few lines do the summation and save the gradient. We pass through every neuron in the visual field of this slice of the current layer. For each neuron, compute k as the ordinal position of this neuron in the complete set. This lets us get the previously computed gradient value in the work area. Recall that d_convgrad_cols is the length of the zero-padded rows of this work area. This causes these fetches to be perfectly coalesced. Note that k could be computed with slightly better efficiency by placing initial computation outside one or both loops. However, this routine requires an insignificant fraction of the total run time, and so clarity is more important. Also note that the store to the gradient is, worst case, very well coalesced.

```
  for (irow=0; irow<d_height[ilayer]; irow++) {
    for (icol=0; icol<d_width[ilayer]; icol++) {
```

```
    k = (irow * d_width[ilayer] + icol) * max_depth + islice; // Neuron at irow, icol, islice
    sum += workptr[k*d_convgrad_cols[ilayer]+iprior];
    }
  }

gradptr[(islice+islice_start)*d_n_prior_weights[ilayer]+iprior] = sum;
}
```

Launch Code for the Gradient

This section presents the code that handles all launches related to computation of the gradient. It contains two complications. First, for convolutional layers, we must deal with the work area. It is allocated during initialization, and this will not be covered here; the complete code can be found in the file MOD_CUDA.cu. However, the code shown here should make clear how the allocation is done.

The second complication is that for any architecture other than tiny, we will break up the task into several launches. There are two reasons for this breakup. First, the memory requirement for the convolutional work area can be large, and its size can be limited by processing subsets of the depth. Second, in most applications, gradient computation is the primary eater of time. By splitting the task into multiple launches, we can prevent the infamous Windows WDDM timeout.

Here is the beginning of this routine:

```
int cuda_hidden_gradient (
  int max_hid_grad,        // Max hid in a CONV hid grad launch
  int max_mem_grad,        // Maximum CONV working memory (MB) per CUDA launch
  int istart,              // Index of first case in this batch
  int nc,                  // Number of cases in batch
  int ilayer,              // Hidden layer being processed
  int type,                // Type of this layer
  int nhid_this,           // Number of hidden neurons in this layer
  int nhid_prior,          // And in prior layer
  int depth,               // Depth of this layer
  int n_prior_weights      // N of inputs per neuron (including bias) to prior layer
  )
```

```
{
  int i, nhid_launch, ihid_start, warpsize, threads_per_block, field, divisor;
  dim3 block_launch;
  cudaError_t error_id;

  field = nhid_this / depth;              // Visual field size = height * width
  warpsize = deviceProp.warpSize;         // Threads per warp, likely 32 into the future
```

The only potentially confusing parameters in the previous calling list are the first two, max_hid_grad and max_mem_grad. They can be set by the user. The first is the maximum number of hidden neurons that may be processed in a launch. Its maximum value is 65535, a concession to device hardware limits. Typically, the user would reduce this in order to bring launch times under the Windows WDDM timeout limit. The second is the maximum number of megabytes of device memory to allocate for a work area for computing the convolutional gradient.

If this is a fully connected layer, we just launch the routine that we saw on page 122. The +1 for threads includes the bias term in the gradient computation.

```
if (type == TYPE_FC) {
  threads_per_block = (nhid_prior + 1 + warpsize - 1) / warpsize * warpsize;
  if (threads_per_block > 4 * warpsize)
    threads_per_block = 4 * warpsize;
  block_launch.x = (nhid_prior + 1 + threads_per_block - 1) / threads_per_block;
  block_launch.y = nhid_this;
  block_launch.z = nc;
  device_hidden_gradient_FC <<< block_launch, threads_per_block >>>
                                    (istart, nc, ilayer);
  cudaDeviceSynchronize();
}
```

The next few lines determine how many hidden neurons will be processed in each of the likely multiple launches.

```
else if (type == TYPE_LOCAL || type == TYPE_CONV) {
  divisor = 1; // Figure out how much we have to divide depth to meet limits

  if (type == TYPE_CONV) {      // For user's scratch memory limitation
    conv_cols = (n_prior_weights + 31) / 32 * 32; // CONV scratch is zero padded
    n_max = 1024 * 1024 * max_mem_grad / (max_batch * conv_cols * sizeof(float));
  }
```

```
else                            // LOCAL layer does not use scratch memory
  n_max = MAXPOSNUM;        // Largest positive number = 2147483647
for (;;) {
  nhid_launch = depth / divisor * field; // We will launch this many hid at a time
  if (nhid_launch <= max_hid_grad && nhid_launch <= n_max)
    break;
  ++divisor;
  }
if (nhid_launch < field)    // Careless user may have set it too small
  nhid_launch = field;      // So ignore it
```

In the previous code, we determine how many (divisor) roughly equally sized launches we need in order to satisfy both of two limits imposed by the user. The user specifies a maximum number of megabytes for the convolution gradient work area. (This is limited to 2,047.) We multiply this by the number of bytes in a megabyte. A single hidden neuron will require max_batch*conv_cols floats, so we divide to get the limit on the number of hidden neurons that can be processed.

Our gradient routine demands that complete visible fields be processed, so trial values of nhid_launch are always a multiple of the field size. We increase the splitting divisor until both user limits are satisfied.

In case a careless user specified a limit so small that at least one visible field cannot be processed, we fix the situation.

The initialization code performed this same operation and allocated the scratch memory according to the largest memory requirement of any layer.

On the next page we show the first half of the launch loop. This loop performs the multiple partial launches, each time processing a multiple of the visible field. The last launch will be smaller than the others if (as is common) division into equal size launches is not possible.

Before starting the launch loop, we zero the convolution work area. This is because the gradient routine will not compute "undefined" entries because of edge padding, but the flattening routine will sum everything. Garbage will wreak havoc. If the final pass is a different size, this zeroing must be repeated.

```
if (type == TYPE_CONV) {
  // We must zero the CONV work area because some entries may be undefined
  // This must also be done in the last pass, because a partial launch at the end
  // may have garbage from the prior launch in 'undefined' locations.
  for (i=0; i<max_convgrad_work; i++)
    fdata[i] = 0.0;    // The gradient routine may leave some of these unset
  error_id = cudaMemcpy (h_convgrad_work, fdata,
              max_convgrad_work * sizeof(float), cudaMemcpyHostToDevice);
}

for (ihid_start=0; ihid_start < depth*field; ihid_start+=nhid_launch) { // Launch loop

  threads_per_block = (n_prior_weights + warpsize - 1) / warpsize * warpsize;
  if (threads_per_block > 4 * warpsize)
    threads_per_block = 4 * warpsize;
  block_launch.x = (n_prior_weights + threads_per_block - 1) / threads_per_block;

  block_launch.y = nhid_launch;
  if (depth*field - ihid_start < nhid_launch) {       // Last launch may be partial
    block_launch.y = depth*field - ihid_start;        // Size of partial launch
    if (type == TYPE_CONV) {                          // Must zero work area again
      for (i=0; i<max_convgrad_work; i++)             // because the layout changed
        fdata[i] = 0.0;
      error_id = cudaMemcpy (h_convgrad_work, fdata,
                  max_convgrad_work * sizeof(float), cudaMemcpyHostToDevice);
    }
  } // If last launch is partial

  block_launch.z = nc;        // Number of cases

  device_hidden_gradient_LOCAL_CONV
        <<< block_launch, threads_per_block >>>
        (type==TYPE_LOCAL ? 1 : 0, n_prior_weights-1, istart,
        ihid_start/field, block_launch.y/field, ilayer);
  cudaDeviceSynchronize();
```

The launch just shown computed the gradient for this set of slices, usually just part of the entire depth of the current layer. If this is a convolutional layer, the individual neuron gradient terms are in the work area. We now need to flatten this matrix by summing across the entire visual field, separately for each layer.

```
if (type == TYPE_CONV) { // Must also flatten gradient?
    threads_per_block = (n_prior_weights + warpsize - 1) / warpsize * warpsize;
    if (threads_per_block > 4 * warpsize)          // It may be sensible to increase
        threads_per_block = 4 * warpsize;          // this limit for modern devices

    block_launch.x = (n_prior_weights + threads_per_block - 1) / threads_per_block;
    block_launch.y /= field;                       // Number of slices in launch
    block_launch.z = nc;                           // Number of cases

    device_flatten_gradient <<< block_launch, threads_per_block >>>
                        (ihid_start / field, block_launch.y, ilayer);
    cudaDeviceSynchronize();
    } // CONV so flatten gradient matrix
  } // Launch loop
 } // LOCAL or CONV

  return 0;
}
```

Fetching the Gradient

The last piece of important CUDA code is the routine for copying the gradient from the device to the host. This happens in two steps. First, a small, simple kernel is launched to sum the individual case gradients into a single gradient for the batch being processed. Then a ridiculously complex routine sums those values into an array in host memory. Why ridiculously complex? Because the order of weights in the device gradient is neither the order of weights on the device nor the order on the host! (Ha! You failed to notice this in the gradient routines, didn't you?) For the device gradient, the input neuron changes fastest, ordered (row, column, slice). The current neuron is also ordered (row, column, slice). It's all about memory coalescing. Most of efficient CUDA programming is, isn't it?

But let's begin with the almost trivial device routine that sums the current batch of case gradients. Each thread is dedicated to a single weight. There is no reason to use a sophisticated summation algorithm like reduction because this routine takes insignificantly small computer time.

```
__global__ void device_fetch_gradient (
  int nc        // Number of cases in batch
  )
{
  int index, icase;
  float *gptr;
  double sum;

  index = blockIdx.x * blockDim.x + threadIdx.x;

  if (index >= d_n_weights)
    return;

  sum = 0.0;
  gptr = d_grad[0] + index;              // Complete gradient starts at [0]
  for (icase=0; icase<nc; icase++)       // For all cases in this batch
    sum += gptr[icase*d_n_weights];
  *gptr = sum;
}
```

Here is the routine called by the host. As it processes batches, it cumulates the sum of the batch gradients in hostgrad. The calling parameters here should all be self-explanatory.

```
int cuda_fetch_gradient (
    int nc,                 // Number of cases in batch
    int n_weights,          // Number of weights
    double **hostgrad,      // Gradient sum output here
    int n_classes,          // Number of outputs
    int n_layers,           // Hidden layers; does not include output
    int *layer_type,        // Type of each layer
    int img_rows,           // Size of input image
    int img_cols,
    int img_bands,
```

```
  int *height,          // Height of visible field in each layer
  int *width,           // Width of visible field
  int *depth,           // Number of slices in each layer
  int *nhid,            // Number of hidden neurons in each layer
  int *hwH,             // Half-width of filters
  int *hwV
  )
{
  int warpsize, blocks_per_grid, threads_per_block;
  int n, n_prior, ilayer, isub, idepth, iheight, iwidth, ndepth, nheight, nwidth;
  int in_row, in_col, in_slice, in_n_height, in_n_width, in_n_depth;
  double *gptr;
  float *fptr;
  cudaError_t error_id;

  warpsize = deviceProp.warpSize;      // Threads per warp, likely 32 well into the future

  threads_per_block = (n_weights + warpsize - 1) / warpsize * warpsize;
  if (threads_per_block > 4 * warpsize)
    threads_per_block = 4 * warpsize;
  blocks_per_grid = (n_weights + threads_per_block - 1) / threads_per_block;

  device_fetch_gradient <<< blocks_per_grid, threads_per_block >>> (nc);
  cudaDeviceSynchronize();

  error_id = cudaMemcpy (fdata, grad, n_weights * sizeof(float),
                         cudaMemcpyDeviceToHost);
```

That much was straightforward. We now have in fdata the sum of individual case gradients for this batch. We will sum them into the host's gradient vector, but they must be reordered.

```
  fptr = fdata;

  for (ilayer=0; ilayer<=n_layers; ilayer++) {
    gptr = hostgrad[ilayer];
```

```
/*
  Fully connected
*/

    if (ilayer == n_layers || layer_type[ilayer] == TYPE_FC) {
      if (ilayer == 0) {
        in_n_height = img_rows;
        in_n_width = img_cols;
        in_n_depth = img_bands;
        }
      else {
        in_n_height = height[ilayer-1];
        in_n_width = width[ilayer-1];
        in_n_depth = depth[ilayer-1];
        }

      n_prior = in_n_height * in_n_width * in_n_depth + 1;
      if (ilayer == n_layers)
        n = n_classes; // Equals depth in fully connected
      else
        n = nhid[ilayer]; // Equals depth in fully connected

      for (idepth=0; idepth<n; idepth++) {
        for (in_row=0; in_row<in_n_height; in_row++) {
          for (in_col=0; in_col<in_n_width; in_col++) {
            for (in_slice=0; in_slice<in_n_depth; in_slice++) {
              // Compute location of this neuron's weight vector in host
              isub = idepth*n_prior + (in_slice*in_n_height + in_row)*in_n_width + in_col;
              gptr[isub] += *fptr++;
              } // For in_slice
            } // For in_col
          } // For in_row
        // Bias
        isub = idepth * n_prior + n_prior - 1;
        gptr[isub] += *fptr++;
        } // For idepth
      }
```

```
/*
  LOCAL
*/

    else if (layer_type[ilayer] == TYPE_LOCAL) {
      // For LOCAL layers, neuron layout in current layer is (height, width, depth).
      n = nhid[ilayer];
      ndepth = depth[ilayer];
      nheight = height[ilayer];
      nwidth = width[ilayer];
      in_n_height = 2 * hwV[ilayer] + 1;
      in_n_width = 2 * hwH[ilayer] + 1;

      if (ilayer == 0)
        in_n_depth = img_bands;
      else
        in_n_depth = depth[ilayer-1];

      n_prior = in_n_height * in_n_width * in_n_depth + 1;

      for (iheight=0; iheight<nheight; iheight++) { // nhid = ndepth * nheight * nwidth
        for (iwidth=0; iwidth<nwidth; iwidth++) {
          for (idepth=0; idepth<ndepth; idepth++) { // Note the order on the dev ice
            for (in_row=0; in_row<in_n_height; in_row++) {
              for (in_col=0; in_col<in_n_width; in_col++) {
                for (in_slice=0; in_slice<in_n_depth; in_slice++) {

                  // Compute location of this neuron's weight in host
                  // First locate the neuron in the current layer, then update per input
                  isub = (idepth * nheight + iheight) * nwidth + iwidth;
                  isub = isub*n_prior+(in_slice*in_n_height + in_row)*in_n_width+in_col;
                  gptr[isub] += *fptr++;
                  } // For in_slice
                } // For in_col
              } // For in_row
            // Bias
            isub = (idepth * nheight + iheight) * nwidth + iwidth; // Neuron in this layer
            isub = isub * n_prior + n_prior - 1;
```

```
          gptr[isub] += *fptr++;
          } // For idepth
        } // For iwidth
      } // For iheight
    }

/*
  CONV
*/

    else if (layer_type[ilayer] == TYPE_CONV) {
      nheight = height[ilayer];
      nwidth = width[ilayer];
      ndepth = depth[ilayer];
      in_n_height = 2 * hwV[ilayer] + 1;
      in_n_width = 2 * hwH[ilayer] + 1;
      if (ilayer == 0)
        in_n_depth = img_bands;
      else
        in_n_depth = depth[ilayer-1];
      n_prior = in_n_height * in_n_width * in_n_depth + 1;
      for (idepth=0; idepth<ndepth; idepth++) { // Just depth; neurons in slice same wts
        for (in_row=0; in_row<in_n_height; in_row++) {
          for (in_col=0; in_col<in_n_width; in_col++) {
            for (in_slice=0; in_slice<in_n_depth; in_slice++) {
              // Compute location of this neuron's weight vector in host
              isub = idepth*n_prior + (in_slice*in_n_height + in_row)*in_n_width + in_col;
              gptr[isub] += *fptr++;
              } // For in_slice
            } // For in_col
          } // For in_row
        //Bias
        isub = idepth * n_prior + n_prior - 1;
        gptr[isub] += *fptr++;
```

```
        } // For idepth
      }
    } // For ilayer
  return 0;
}
```

Putting It All Together

We've seen most of the individual components of gradient computation. We finish this CUDA chapter with the host routine that calls the routines presented to this point. Here is its beginning. The caller of this routine can specify via jstart and jstop a range of cases in the training set to be processed. This facilitates advanced training/testing algorithms. The caller also specifies whether this routine is to compute the gradient in addition to the performance criterion.

```
double Model::model_cuda (int find_grad, int jstart, int jstop)
{
  int i, nc, ilayer, ret_val, ibatch, n_in_batch, n_subsets, max_batch, istart, istop;
  int n_done, n_launches, n_prior, ineuron, ivar;
  double ll, *wptr, *gptr, wt, wpen;

  nc = jstop - jstart;         // Number of training cases to process
```

To prevent integer overflow in allocating memory for the gradient, we compute the minimum number of subsets (n_subsets) needed to get each subset small enough. Here, max_batch is the maximum batch size (number of cases in a batch). The CUDA initialization call will allocate max_batch * n_all_weights floats. The unit of execution is a single case, so we will compute the gradient requirement of each individual case. Recall that the model member variable n_all_weights is the total number of weights for the model, and MAXPOSNUM is the maximum positive number. We could do some fancier math using unsigned integers, but that's tricky and fraught with opportunities for error. Plus, this limit will not often be hit, and smaller batches are fine anyway.

```
  max_batch = MAXPOSNUM / (n_all_weights * sizeof(float)); // Memory allocation size
  if (max_batch > 65535)                   // Grid dimension hardware limitation
    max_batch = 65535;
```

```
// The user may want to split into more subsets to prevent CUDA timeout
if (max_batch > TrainParams.max_batch)
  max_batch = TrainParams.max_batch;

n_subsets = (nc + max_batch - 1) / max_batch;
```

The CUDA device must be initialized once. In unusual situations, the actual maximum batch size may be a little different from that computed previously (but still safe), so now that we know the number of subsets, we recompute the max batch size once again, just to be sure. The following little loop is exactly the same form as that which will control division of the training set into separately processed subsets. For each batch, it computes the number of cases to do in this batch by looking at the number left to do and dividing by the number of batches left to do. Then it calls the CUDA initialization routine, which allocates memory on the device, initializes local constants, and so forth. The complete source code for this routine is in the file MOD_CUDA.cu.

```
if (! cuda_initialized) {
  n_done = 0;        // Must find max batch size for cuda init
  for (ibatch=0; ibatch<n_subsets; ibatch++) {
    n_in_batch = (nc - n_done) / (n_subsets - ibatch);    // Cases left to do / batches left
    if (ibatch == 0 || n_in_batch > max_batch)
      max_batch = n_in_batch;
    n_done += n_in_batch;
    }
  cuda_init (...);
  cuda_initialized = 1;
  }
```

There is a global variable called cuda_weights_changed. It is set to one any time the model weights are adjusted by the training routine. Then, when this criterion/gradient routine is called, it checks this variable and sends the new set of weights to the device if the flag is set.

```
if (cuda_weights_changed) {
  ret_val = cuda_weights_to_device (...);
  cuda_weights_changed = 0;
  }
```

We will sum the gradient across all batches, so it must be zeroed.

```
if (find_grad) {
  for (i=0; i<n_all_weights; i++)
    gradient[i] = 0.0;
  }
```

The main batch loop now begins. The user's starting and stopping cases are in jstart and jstop. We break this range into batches with istart and istop.

```
istart = jstart;
n_done = 0;        // Number of training cases done in this epoch so far

for (ibatch=0; ibatch<n_subsets; ibatch++) {
  n_in_batch = (nc - n_done) / (n_subsets - ibatch);      // Cases left / batches left
  istop = istart + n_in_batch;                            // Stop just before this
```

```
/*
  Forward pass
*/
```

```
    for (ilayer=0; ilayer<n_layers; ilayer++) {        // All hidden; do output separately
      if (layer_type[ilayer] == TYPE_FC)
        ret_val = cuda_hidden_activation_FC (istart, istop, nhid[ilayer], ilayer);
      else if (layer_type[ilayer] == TYPE_LOCAL)
        ret_val = cuda_hidden_activation_LOCAL_CONV_shared (1, istart, istop,
              nhid[ilayer], depth[ilayer], ilayer);
      else if (layer_type[ilayer] == TYPE_CONV)
        ret_val = cuda_hidden_activation_LOCAL_CONV_shared (0, istart, istop,
              nhid[ilayer], depth[ilayer], ilayer);
      else if (layer_type[ilayer] == TYPE_POOLAVG)
        ret_val = cuda_hidden_activation_POOLED (1, istart, istop, nhid[ilayer],
                                              depth[ilayer], ilayer);
      else if (layer_type[ilayer] == TYPE_POOLMAX)
        ret_val = cuda_hidden_activation_POOLED (0, istart, istop, nhid[ilayer],
                                              depth[ilayer], ilayer);
      } // For ilayer
```

```
/*
  Output layer going forward, then SoftMax
*/

    if (n_layers == 0)
      ret_val = cuda_output_activation_no_hidden (istart, istop);
    else
      ret_val = cuda_output_activation (istart, istop);
    ret_val = cuda_softmax (istart, istop);
```

The previous code loops through all hidden layers (but not the output layer) in a forward pass, computing activations layer by layer. Then it computes the output activation, using separate routines depending on if there are hidden layers versus direct input-to-output connections. Finally, it does the SoftMax conversion of outputs.

If the caller also wants the gradient, we zero the gradient on the device because for some (poorly designed) models, some terms may truly be zero but are architecturally undefined. Compute the output delta and gradient and then loop backward through all hidden layers, backpropagating delta and computing the gradient.

```
    if (find_grad) {

      ret_val = cuda_zero_gradient (istop-istart, n_all_weights);
      ret_val = cuda_output_delta (istart, istop, n_classes);
      if (n_layers == 0)
        ret_val = cuda_output_gradient (istart, istop-istart, n_pred, -1, n_classes);
      else
        ret_val = cuda_output_gradient (istart, istop-istart, nhid[n_layers-1],
                                        n_layers-1, n_classes);

      for (ilayer=n_layers-1; ilayer>=0; ilayer--) {

        if (ilayer == n_layers-1 || layer_type[ilayer+1] == TYPE_FC)
          ret_val = cuda_backprop_delta_FC (istop-istart, ilayer, nhid[ilayer]);
        else if (layer_type[ilayer+1] == TYPE_LOCAL ||
                 layer_type[ilayer+1] == TYPE_CONV)
          ret_val = cuda_backprop_delta_nonpooled (istop-istart, ilayer, nhid[ilayer]);
        else if (layer_type[ilayer+1] == TYPE_POOLAVG ||
                 layer_type[ilayer+1] == TYPE_POOLMAX)
          ret_val = cuda_backprop_delta_pooled (istop-istart, ilayer, nhid[ilayer]);
```

```
    ret_val = cuda_move_delta (istop-istart, nhid[ilayer]); // Move prior to this

    ret_val = cuda_hidden_gradient (TrainParams.max_hid_grad,
                  TrainParams.max_mem_grad, istart, istop-istart, ilayer,
                  layer_type[ilayer], nhid[ilayer], ilayer ? nhid[ilayer-1] : n_pred,
                  depth[ilayer], n_prior_weights[ilayer], &n_launches);
    } // For all layers, going backwards
```

After the backward passes are complete, we fetch the gradient, adding it into our batch sum, and then loop back for the next batch. When all batches have been processed, we sum the log likelihood criterion across all training cases and normalize the gradient in the same way we normalize the criterion. The final step before returning is to apply the weight penalty, but this code will not be shown here, as it is long and identical to what we saw on page 59.

```
    ret_val = cuda_fetch_gradient (istop-istart, n_all_weights, layer_gradient,
                      n_classes, n_layers, layer_type,
                      IMAGE_rows, IMAGE_cols, IMAGE_bands,
                      height, width, depth, nhid,
                      HalfWidH, HalfWidV);
      } // If find_grad

    n_done += n_in_batch;
    istart = istop;              // Advance to the next batch
      } // For ibatch

  ret_val = cuda_ll (nc, &ll);

  if (find_grad) {
    for (i=0; i<n_all_weights; i++)
      gradient[i] /= (nc * n_classes);
    }

... Apply weight penalty ...

  return ll / (nc * n_classes) + penalty; // Negative log likelihood
}
```

CHAPTER 4

CONVNET Manual

This chapter is a user's manual for the CONVNET program, available as a free download from my web site. The first section lists every menu option, along with a brief description of its purpose and the page number on which more details can be found if the short description is not sufficient.

Menu Options

First we'll look at the menu options.

File Menu

These are the options on the File menu.

Read control file, page 150

A standard text file is read. This file contains architectural specifications for the model (this is the only way to define architecture) and optionally may contain commands to read or create input images or train the model.

Read MNIST image

A standard MNIST-format file is read. The corresponding label file must be read after the image file is read. Only one MNIST image/label pair may be read. Other file reading options are disabled after an MNIST image/label pair is read. It is assumed that there will be ten classes; this is hard-coded into the program. However, the size of the images is not hard-coded. It is read from the file. The product of the number of rows times the number of columns

© Timothy Masters 2018
T. Masters, *Deep Belief Nets in C++ and CUDA C: Volume 3*, https://doi.org/10.1007/978-1-4842-3721-2_4

cannot exceed 2^16–1=65,535. This unfortunate limitation comes from a hardware property of current CUDA devices, which would be difficult to work around.

Read MNIST labels

A standard MNIST-format label file is read. It is assumed that there are ten classes. The corresponding MNIST image file must be read before the label file is read.

Read CIFAR-10 image

A standard CIFAR-10-format file is read. Multiple CIFAR-10 files may be read, in which case they are concatenated. This command cannot be used if MNIST or series data is already present.

Read series, page 151

A univariate time series is read, and a set of predictors is computed based on the values of the series, optionally differenced and/or log transformed. Class identities are generated. This selection brings up a menu in which parameters relevant to reading the series may be entered. These parameters, in the context of control files, are discussed starting on page 151.

Make image

An artificial image having random tones is generated to enable quick and easy testing of data and model configurations. The user specifies the height and width, the number of bands, the number of classes, and the number of cases. This command cannot be used if a dataset is already present.

Clear all data

All training data is erased, but a trained model (if it exists) is retained. The purpose of this command is to allow reading a test dataset and evaluating the performance of a trained model on this new dataset. A common sequence of operations is ***Read training data, Train, Clear, Read test data, Test***.

Print

> The currently selected display window (created under the Display menu) is printed. If no window is selected, Print is disabled.

Exit

> The program is terminated.

Test Menu

These are the options on the Test menu.

Use CUDA (Toggle Yes/No)

> This option is enabled only if a CUDA-capable device is present on the computer. If a check mark appears next to this option, the CUDA device will be used for compute-intensive operations. Click this option to toggle the check mark on and off.

Training params, page 156

> Parameters relevant to training can be set. This selection brings up a dialog box in which these parameters may be changed from their default values. The nature of these parameters is discussed in the context of a control file on page 156.

Train, page 159

> The model is trained using the data currently present. It is important to understand which phases of training can and cannot be interrupted with the Escape key. See page 159 for details.

Test

> The trained model is tested with the data currently present. The current version of CONVNET does not allow interruption of computing the confusion matrix; you'll just have to sit and wait for it to finish. Sorry. It's on my list, but for some technical reasons it's not a quick-and-easy fix. I hope to post updated versions of the program on my web site as improvements occur.

Print model weights

All model weights are printed to the CONVNET.LOG file. This can be gigantic! Even modest models can have so many weights that writing them to the CONVNET.LOG file can take several minutes and consume megabytes. You've been warned.

Display Menu

These are the options on the Display menu.

Display training images, page 160

A user-selectable set of the images in the current dataset is displayed.

Display filter images, page 160

If a trained model exists and the first hidden layer of this model is convolutional, this option displays as images the filter weights for a user-selectable set of slices.

Display activation images, page 161

If a trained model exists, this option displays as images the activations of the visual field of the first hidden layer for a user-selectable set of slices and training case.

Read Control File

Intelligent readers will study this section and learn to perform most or all operations via a control file. Every CONVNET operation except specifying the model architecture can be done with the menu system, which may be the preferable approach if one is just idly fooling around. However, in the vast majority of cases, it is best for the user to first create a control file using any ordinary text editor and completely specify all project details in this file. This avoids tedious repetitive entry of parameters via the menu system, and it also provides hard documentation of all project specifications.

A control file is an ordinary text file. Each line of this file specifies a single aspect of the project. Comments can be inserted by starting a line with two forward slashes (//). This also provides a convenient mechanism for temporarily deactivating lines in the file without deleting them.

Making and Reading Image Data

This section describes methods for making random test images as well as reading popular-format image files.

MAKE IMAGE Rows Columns Bands Classes Cases

This produces a set of training images having random tones. The user specifies the height and width, the number of bands, the number of classes, and the number of cases. This command cannot be used if a dataset is already present.

READ MNIST IMAGE "FileName"

An MNIST image file is read. This command cannot be used if a dataset is already present. The corresponding label file must be read after the image file is read.

READ MNIST LABELS "FileName"

An MNIST label file is read. This command would normally follow a READ MNIST IMAGE command.

READ C10 IMAGE "FileName"

A CIFAR-10 image file is read. This command cannot be used if a dataset other than CIFAR-10 is already present. Multiple CIFAR-10 image files may be read, and their contents will be concatenated.

Reading a Time Series as Images

This is a powerful technique for converting a time series to a set of images. A moving window is passed across a time series. Each placement defines an image. This window image is divided into a user-specified number of rows (value of the series) and columns (relative time in the window). The path of the series is set to black in the image, and everything else is set to white. Figure 4-1 shows a typical set of images produced from prices of OEX, the Standard and Poor's 100 index, as the window slides along left to right.

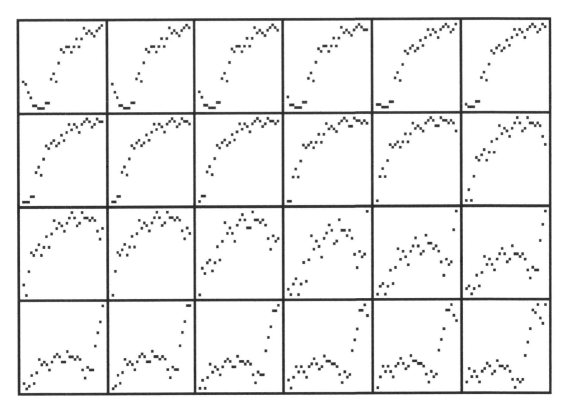

Figure 4-1. *Series images from OEX*

The command to read the series and produce the image set is shown below. Nothing else need be specified. However, in most cases users will want to change some specifications from their defaults. The legal specifications are also shown, with their default values indicated. Naturally, all such specifications must appear *before* the READ SERIES command to which they will apply.

The series file must be an ordinary text file. It may contain a header, and it may contain multiple columns. If there are multiple columns, then spaces, tabs, and commas serve as delimiters. There is one observation per record.

READ SERIES "FileName"

> A time series file is read. This command cannot be used if a
> dataset is already present. A moving window is applied to the
> series to produce a set of images.

SERIES COLUMN = Column

The series data can be fetched from any column. This specifies the column containing the desired values. The default is 1.

SERIES WINDOW = Width

This is the number of records in each window placement. Hence, it is the width of the images. The default is 16.

SERIES RESOLUTION = Resolution

This is the vertical resolution in each window placement. Hence, it is the height of the images. The default is 16.

SERIES SHIFT = Shift

This is the number of records that each window placement will advance to produce the next image. The default is 1.

SERIES RAWDATA

This, the default, specifies that the values read from the file are used as the series data.

SERIES RAWLOG

This specifies that the log of the values read from the file are used as the series data.

SERIES DIFFDATA

This specifies that the differences in the values read from the file are used as the series data. In other words, each computed series value is the current value of the file series minus the prior value.

SERIES DIFFLOG

This is identical to SERIES DIFFDATA except that the difference of the logs is used. Equivalently, this is the log of the ratios.

SERIES FRAC FULL = Fraction

This is the fraction (0–1) of training set cases that are forced to occupy the full vertical range of the window. Windows are not necessarily individually normalized (scaled), as this would distort

153

information content. Normalization is usually relative to the entire series. A specification of zero maps the greatest range of the series across all windows to the full vertical range of the window, meaning that (except for ties) only one window will display the full vertical range. In many situations this will result in many or most windows having very little variation; they are essentially a flat line. A specification of one causes each window to be individually normalized, so all windows display the full vertical range. This is probably not good, as it fails to distinguish windows having little series variation from those having great variation; that's important information, and it's lost. The default is 0.2. This means that the 80th percentile (1 minus 0.2) of within-window ranges is the variation that maps to the full vertical range for those 80 percent of cases. The 20 percent of windows whose series range exceeds this quantity are individually normalized to full vertical range. A simple way of thinking about this specification is that this is the fraction of cases that are individually normalized to the full vertical range. In most applications this should be well under 0.5.

SERIES TARGET NO DIFF

This, the default, specifies that the target class is determined by the next value in the series past the window. This determination will be based on the undifferenced or differenced nature of the series. In other words, the target will be determined by the difference between the next value outside the window minus the last value in the window, if and only if the user specifies that the series is differenced. *Differencing of the target matches the predictors.*

SERIES TARGET DIFF

Specify this option if the series is not differenced (RAWDATA or RAWLOG) but you want the target determination to be based on differences. This would be appropriate, for example, in financial market prediction.

SERIES CLASS ZERO

>This, the default, specifies that the class of a case is defined by the sign of the target (which may or may not have been differenced, as earlier). One class is for targets greater than zero, and the other is for targets less than or equal to zero.

SERIES CLASS MEDIAN

>This specifies that the class of a case is defined by the value of the target relative to the median across the training set. One class is for targets greater than the median, and the other is for targets less than or equal to the median.

SERIES CLASS THIRDS

>This specifies that the class of a case is defined by the value of the target relative to the 33rd and 66th percentiles across the training set. There are three classes: low, middle, and high.

SERIES NO HEADER

>This, the default, specifies that the series file has no header record. The data begins with the first record.

SERIES HEADER

>This specifies that the series file has a header, so the first record is skipped.

Model Architecture

The architecture of the model must be specified in a control file; there is no menu interface for doing so. Layers of the model are given in order from the first hidden layer to the last. There are no specifications for the input and output layers. The following layer types may be defined:

FULLY CONNECTED LAYER *Slices*

>This creates a fully connected layer consisting of the specified number of slices. In architecture reports, it will appear as having one row, one column, and a depth equal to the number of slices.

LOCAL LAYER Slices hwV hwH padV padH strideV strideH

> This creates a locally connected layer having the specified number of slices, vertical and horizontal half-widths, vertical and horizontal padding, and vertical and horizontal stride. The dimensions of the visual field of this layer are given by Equation 1-8.

CONVOLUTIONAL LAYER Slices hwV hwH padV padH strideV strideH

> This creates a convolutional layer having the specified number of slices, vertical and horizontal half-widths, vertical and horizontal padding, and vertical and horizontal stride. The dimensions of the visual field of this layer are given by Equation 1-8.

POOLED AVERAGE LAYER widthV widthH strideV strideH

> This creates an average pooling layer with the specified vertical and horizontal widths (not half-widths) and stride. The dimensions of the visual field of this layer are given by Equation 1-8. The number of slices is equal to the number in the prior layer.

POOLED MAX LAYER widthV widthH strideV strideH

> This creates a max pooling layer with the specified vertical and horizontal widths (not half-widths) and stride. The dimensions of the visual field of this layer are given by Equation 1-8. The number of slices is equal to the number in the prior layer.

Training Parameters

The following parameters relevant to training may be set. Default values are as indicated. It may be that a revised CONVNET program may change these defaults from those that are printed here. The defaults for the current version of the program can be seen by selecting the Test ➤ Training parameters menu option.

MAX BATCH = Number

> This is relevant only for CUDA training. Kernel launches are divided into subsets of the full training set to prevent the infamous Windows WDDM timeout. This parameter limits the maximum number of cases in a subset. The default is 100. Lower this number to lower the per-launch time for all training steps.

MAX HID GRAD = Number

This is the maximum number of hidden neurons that will be processed per launch during CUDA gradient computation of convolutional and locally connected layers. Lowering this number can reduce the per-launch time for gradient computation, without affecting any other aspect of training. In many situations, it is best to leave this be huge and limit the time with the next parameter, MAX MEM GRAD. The default is 65535, which is the maximum legal value.

MAX MEM GRAD = Number

This is the preferred way to lower the time required for gradient computation of convolutional and locally connected layers. It does not impact any other operations. This specifies the maximum memory in megabytes to dedicate to scratch work for convolutional hidden layers. A useful side effect is that limiting the memory causes launches to be broken into smaller sets of hidden neurons, which reduces the per-launch compute time and hence can prevent Windows WDDM timeouts. Lower this number to reduce per-launch compute time. You may also want to use a smaller number if your CUDA device has limited onboard memory. The default is 2047 megabytes, which is the maximum legal value.

To summarize the prior three parameters, Windows limits CUDA computation time for a single kernel launch. The limit is generally two seconds. If this time is exceeded, the screen will temporarily go black, an error message will appear, and the application will be severely compromised. If this happens, you must reduce per-kernel time. Study the CUDA.LOG file to see where excessive per-launch time is occurring. Activation and gradient computation are the only serious time eaters. The MAX BATCH parameter impacts all operations. The MAX HID GRAD and MAX MEM GRAD parameters affect only gradient computation for locally connected and convolutional layers. Adjust these three parameters as needed to bring per-launch time under the Windows limit. The default values apply no limitation, which is good whenever possible, as breaking the task into multiple launches introduces significant overhead.

ANNEAL ITERS = Number

This is the number of simulated annealing iterations used to find good starting weights for refinement. The user can interrupt annealing by pressing the Escape key, at which point refinement will commence with the best weights found so far. The default is 100.

ANNEAL RANGE = Number

This is the approximate range of random values tried at the start of simulated annealing. Larger values provide a wider search space but are also more likely to produce excessively large initial weights that can never be reduced to reasonable values. It's better to err on the side of too small than too large. The default is 0.1.

MAX ITERS = Number

This is the maximum number of conjugate gradient iterations used for weight refinement. The default is 1000. It may be good to set this to a smaller value if you are processing a collection of training operations in a single control file. However, in most cases it's best to make this a very large number and use the next parameter, *TOL*, to end training. Or you can manually interrupt training when the criterion graph looks like it has stabilized.

TOL = Number

This is the preferred method for determining convergence of the weight refinement algorithm. Roughly speaking, this specifies the degree of iteration-to-iteration criterion improvement for deciding that convergence is obtained. The default is 0.00005. Smaller values will force more extended training. Training ends when either MAX ITERS or TOL is hit.

WPEN = Number

This is the weight penalty, which penalizes large weights. A positive value will, by definition, degrade the performance criterion of the trained model. However, because large weights are often associated with overfitting, one may obtain better out-of-sample performance. The default is zero. A little weight penalty goes a long way, so if you experiment, start out very small, such as 0.001 or so.

Operations

As of now, there are three operations that can be performed with CONVNET within a control file. These are as follows:

TRAIN

> A model is trained using the current dataset. This operation can be roughly divided into four phases. In the first phase, simulated annealing is used to find good starting weights for subsequent refinement. Pressing the Escape key interrupts annealing, and refinement will proceed with the best weights found so far.

> The second phase is weight refinement using conjugate gradient optimization. This, too, can be interrupted with Escape. However, in some cases the computer may take considerable time to respond, as certain subphases are not interruptible. Be patient.

> The third phase is short, a final pass through the data with the best weights found. This phase can be interrupted with Escape. However, doing so will cause all results to be lost. Be warned.

> The fourth phase is computation of the confusion matrix. Unfortunately, the current version of CONVNET does not allow interruption of this operation. Patience is a virtue.

TEST

> This assumes that a dataset and a trained model are present. Performance criteria, mainly the confusion matrix, are computed.

CLEAR

> All data is erased, but a trained model, if present, is not disturbed. The usual purpose of this command is to allow reading of a test set after a model has been trained. The usual sequence is as follows:

> ***Read training data***

> ***Train***

> ***Clear***

> ***Read test data***

> ***Test***

Display Options

Several options for displaying useful information as screen images are available. They are described in this section.

Display Training Images

Images from the training set are displayed. This option is enabled only if the images have one or three bands. The user enters the following information on a menu:

First to display

> This is the ordinal number (1 is the first) of the first training set case to display. Images start in the upper-left corner of the screen and advance left to right first. If the total number to display exceeds the number in the training set, cases will wrap around to the first case in the training set.

Rows

> This many rows of images will be displayed.

Columns

> This many columns of images will be displayed. The total number of training cases displayed is *Rows* times *Columns*.

Display Filter Images

If the input image has either one or three bands, a trained model exists, and the first hidden layer of this model is convolutional, this option displays filter weights as images. The displayed images have the same dimensions and orientation as the filter.

If the input image has one band, the display is black and white, with strongly negative weights being black and strongly positive weights being white. Intermediate weights are shades of gray.

If the input image has three bands, the display uses a three-color display, with red, green, and blue matching the corresponding colors in the input image. For example, if the weights for all three bands are strongly negative, the corresponding image pixel will be black. If all three are strongly positive, the pixel will be white. A red pixel means that the weight for the red channel of the input image is strongly positive, and the weights for the other two channels are strongly negative. Et cetera. The user specifies the following parameters:

First slice to display

> This is the ordinal number of the first slice to display. Images start in the upper-left corner of the screen and advance left to right first. If the total number to display exceeds the number of slices, they will wrap around to the first slice.

Rows for slices

> This many rows of slice images will be displayed.

Columns for slices

> This many columns of slice images will be displayed. The total number of slices displayed is *Rows* times *Columns*.

Scale slices individually

> By default, the scale for mapping weights to tone is determined by examining all *Rows* times *Columns* displayed weights. If this box is checked, scaling is applied to each image separately, which may over-emphasize low-utility filters.

Display Activation Images

If a trained model and dataset are present, we can display the activations of the first hidden layer (any layer type) as images. The images are black and white, with black representing the lowest activation possible, and white the highest.

The user specifies the following parameters:

First slice to display

> This is the ordinal number of the first slice to display. Images start in the upper-left corner of the screen and advance left to right first. If the total number to display exceeds the number of slices, they will wrap around to the first slice.

Rows for slices

> This many rows of slice images will be displayed.

Columns for slices

> This many columns of slice images will be displayed. The total number of slices displayed is *Rows* times *Columns*.

Case number

> This is the ordinal number of the training case whose activations are displayed. It must not exceed the number of training cases.

Example of Displays

This section provides an example demonstrating the several display options that are available.

Figure 4-2 shows an example of the numeral zero taken from the MNIST dataset. A model consisting of a single convolutional layer having eight slices is created to train using the MNIST dataset. Figure 4-3 shows what the weights for each of these eight slices look like early in the training process. Note the great randomness. Figure 4-4 shows the same display after training has progressed to convergence. Note how clear response patterns have emerged. Finally, Figure 4-5 shows the activation pattern of the eight slices when presented with the MNIST zero of Figure 4-2.

It's worth pursuing this a little further. Look at the weight pattern in the second slice (top row, second from left) of Figure 4-4. It's very bright (high positive weights) near the center, and it's fairly or greatly dark (zero or negative weights) elsewhere. As one would expect, the activation pattern for the same slice in Figure 4-5 largely replicates the input image, though with some blurring.

Compare this with the last (bottom-right) slice. This weight set is just the opposite, being very dark (negative weights) in the center. We see in the corresponding activation display that the pattern is the negative of the input image. Lovely.

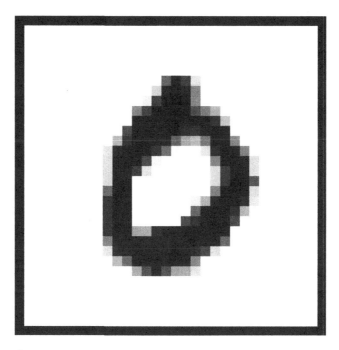

Figure 4-2. *MNIST zero*

Figure 4-3. *Weights early in training*

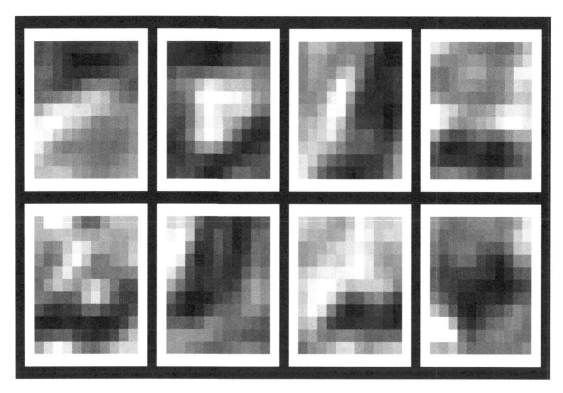

Figure 4-4. *MNIST weights trained to convergence*

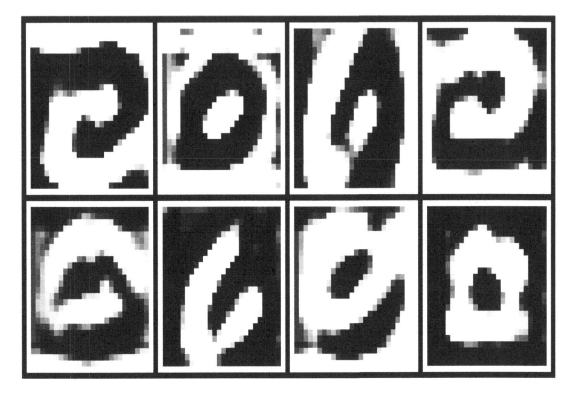

Figure 4-5. *MNIST zero activations*

The CONVNET.LOG file

The CONVNET program writes a log file that contains information about all operations. To understand this file, the following control file was created. It employs every available layer type.

```
MAKE IMAGE 12 12 1 6 1024
CONVOLUTIONAL LAYER 6 1 1 1 1 1 1
POOLED MAX LAYER 3 3 2 2
LOCAL LAYER 3 1 1 1 1 1 1
POOLED AVERAGE LAYER 3 3 2 2
FULLY CONNECTED LAYER 4
WPEN = 0.001
TRAIN
```

The log file echoes these lines, which we will skip here. The first important section in the log file is its description of the model's architecture.

Input has 12 rows, 12 columns, and 1 bands

Model architecture...

Model has 6 layers, including fully connected output

  Layer 1 is convolutional, with 6 slices, each 12 high and 12 wide

    Horz half-width=1, padding=1, stride=1

    Vert half-width=1, padding=1, stride=1

    864 neurons and 10 prior weights per slice gives 60 weights

  Layer 2 is 3 by 3 pooling max, with stride 2 by 2, 5 high, 5 wide, and 6 deep

  Layer 3 is locally connected, with 3 slices, each 5 high and 5 wide

    Horz half-width=1, padding=1, stride=1

    Vert half-width=1, padding=1, stride=1

    75 neurons and 55 prior weights per neuron gives 4125 weights

  Layer 4 is 3 by 3 pooling average, with stride 2 by 2, 2 high, 2 wide, and 3 deep

  Layer 5 is fully connected, with 4 slices, each 1 high and 1 wide

    4 neurons and 13 prior weights per neuron gives 52 weights

  Layer 6 (output) is fully connected, with 6 slices (classes)

    6 neurons and 5 prior weights per neuron gives 30 weights

  4267 Total weights for the entire model

Because the first layer (convolutional) has the padding equal to the half-width and no striding, we see that it has the same visual field dimensions as the input layer. If necessary, review Equation 1-8. The layer has $12*12*6=864$ neurons. The filter size is $((2*1+1)^2)*1+1=10$. (The *1 is the depth of the prior layer, and the +1 is the bias term.) All neurons in the visual field share the same weight set, so the total number of weights for the layer is the filter size (10) times the number of slices (6).

Equation 1-8 gives the size of the second layer: $(12-3+0)/2+1=5$.

Layer 3 has the padding equal to the half-width and no striding, so its visual field dimensions are the same as the prior layer. The filter size is $((2*1+1)^2)*6+1=55$. There is a different weight set for each of the $5*5*3=75$ neurons in this layer, giving a total of 4125 weights for this layer.

Equation 1-8 gives the size of the fourth layer: $(5-3+0)/2+1=2$.

Layer 5 is fed by $2*2*3$ neurons in the prior layer. Including the bias term gives 13 weights per neuron. This layer has 4 neurons, so it has a total of 52 weights. Recall that our convention is that fully connected layers have a $1*1$ visual field, with a depth equal to the number of neurons.

Layer 6, the output layer, is by definition fully connected. It's fed by 1*1*4 neurons in the prior layer. Including the bias gives 5 weights per neuron. It has a depth of 6, the number of classes, so it has 30 weights.

Adding these gives a total of 4,267 weights in the model.

Simulated annealing completes, but I interrupted refinement. The following lines appear:

Simulated annealing for starting weights is complete with mean negative log likelihood = 0.29804

WARNING... User pressed ESCape during optimization
        Results are incomplete and may be seriously incorrect

Optimization is complete with negative log likelihood = 0.09214

The last item printed is a confusion matrix. The row (in groups of three) is the true class, and the column is the predicted class. In each set of three rows for a true class, the first row is the count, the second row is the percent for that row (true class), and the third row is the percent of the entire dataset.

	1	2	3	4	5	6
1	168	0	2	0	2	0
	97.67	0.00	1.16	0.00	1.16	0.00
	16.41	0.00	0.20	0.00	0.20	0.00
2	1	127	18	0	1	31
	0.56	71.35	10.11	0.00	0.56	17.42
	0.10	12.40	1.76	0.00	0.10	3.03
3	1	12	120	2	4	11
	0.67	8.00	80.00	1.33	2.67	7.33
	0.10	1.17	11.72	0.20	0.39	1.07
4	8	0	1	124	48	1
	4.40	0.00	0.55	68.13	26.37	0.55
	0.78	0.00	0.10	12.11	4.69	0.10
5	6	1	0	11	178	0
	3.06	0.51	0.00	5.61	90.82	0.00
	0.59	0.10	0.00	1.07	17.38	0.00
6	0	20	1	0	1	124
	0.00	13.70	0.68	0.00	0.68	84.93
	0.00	1.95	0.10	0.00	0.10	12.11

Total misclassification = 17.8711 percent

Printed Weights

The user has the option of printing weights for the entire model. Be warned that the total number of weights can be enormous, in which case the resulting file will also be enormous, and it may even require several minutes of run time to do the file writing. Here is a partial listing of the weights for the example cited in the prior section. Please reconcile this listing with the architecture of this model.

```
Layer 1 of 6 (Convolutional)  Slice 1 of 6
       3.642629  Input band 1 Neuron 1
      -0.676231  Input band 1 Neuron 2
      -0.085785  Input band 1 Neuron 3
       2.766258  Input band 1 Neuron 4
      -2.646048  Input band 1 Neuron 5
      -0.865142  Input band 1 Neuron 6
       1.900750  Input band 1 Neuron 7
      -2.298438  Input band 1 Neuron 8
       0.924283  Input band 1 Neuron 9
      ----------------------------------
      -3.971506  BIAS

... (Slices 2-5)

Layer 1 of 6 (Convolutional)  Slice 6 of 6
       3.011171  Input band 1 Neuron 1
       0.687377  Input band 1 Neuron 2
       1.019491  Input band 1 Neuron 3
      -0.832090  Input band 1 Neuron 4
       1.724954  Input band 1 Neuron 5
      -1.247742  Input band 1 Neuron 6
       0.444635  Input band 1 Neuron 7
       1.737460  Input band 1 Neuron 8
      -0.542140  Input band 1 Neuron 9
      ----------------------------------
      -2.507262  BIAS

Layer 2 of 6 (Mean pool) 5 rows by 5 cols by 6 slices
```

Layer 3 of 6 (Local)Slice 1 of 3 Row 1 of 5 Col 1 of 5

     0.016978 Prior layer slice 1 Neuron 1

    -0.027422 Prior layer slice 1 Neuron 2

    -0.052678 Prior layer slice 1 Neuron 3

     0.036557 Prior layer slice 1 Neuron 4

    -0.755227 Prior layer slice 1 Neuron 5

     0.211502 Prior layer slice 1 Neuron 6

     0.036439 Prior layer slice 1 Neuron 7

    -0.398360 Prior layer slice 1 Neuron 8

     0.737985 Prior layer slice 1 Neuron 9

... Other rows and columns, then slice 2 and part of 3

Layer 3 of 6 (Local)Slice 3 of 3 Row 5 of 5 Col 5 of 5

    -1.035432 Prior layer slice 1 Neuron 1

    -0.357207 Prior layer slice 1 Neuron 2

    -0.021757 Prior layer slice 1 Neuron 3

    -0.033135 Prior layer slice 1 Neuron 4

    -0.107814 Prior layer slice 1 Neuron 5

    -0.000594 Prior layer slice 1 Neuron 6

    -0.051112 Prior layer slice 1 Neuron 7

     0.023901 Prior layer slice 1 Neuron 8

    -0.020555 Prior layer slice 1 Neuron 9

... Slices 2 through 5

     0.679523 Prior layer slice 6 Neuron 1

    -1.053021 Prior layer slice 6 Neuron 2

     0.001994 Prior layer slice 6 Neuron 3

    -0.104741 Prior layer slice 6 Neuron 4

    -0.664431 Prior layer slice 6 Neuron 5

     0.034758 Prior layer slice 6 Neuron 6

     0.016724 Prior layer slice 6 Neuron 7

     0.014839 Prior layer slice 6 Neuron 8

     0.050983 Prior layer slice 6 Neuron 9

    -1.963063 BIAS

Layer 4 of 6 (Avg pool) 2 rows by 2 cols by 3 slices

Layer 5 of 6 (Full) Slice (this neuron) 1 of 4
      1.592443 Prior layer slice 1 Neuron 1
      1.161122 Prior layer slice 1 Neuron 2
    -0.162907 Prior layer slice 1 Neuron 3
     0.648188 Prior layer slice 1 Neuron 4
    -1.275991 Prior layer slice 2 Neuron 1
    -3.782788 Prior layer slice 2 Neuron 2
    -2.344005 Prior layer slice 2 Neuron 3
    -2.019643 Prior layer slice 2 Neuron 4
    -0.240221 Prior layer slice 3 Neuron 1
    -0.118739 Prior layer slice 3 Neuron 2
     0.739422 Prior layer slice 3 Neuron 3
     1.031370 Prior layer slice 3 Neuron 4
    -0.878146 BIAS

...

Layer 5 of 6 (Full) Slice (this neuron) 4 of 4
     0.560776 Prior layer slice 1 Neuron 1
    -0.467746 Prior layer slice 1 Neuron 2
    -1.281872 Prior layer slice 1 Neuron 3
    -0.444215 Prior layer slice 1 Neuron 4
     0.948946 Prior layer slice 2 Neuron 1
     1.805807 Prior layer slice 2 Neuron 2
     1.796881 Prior layer slice 2 Neuron 3
     1.776497 Prior layer slice 2 Neuron 4
     4.415077 Prior layer slice 3 Neuron 1
     2.461983 Prior layer slice 3 Neuron 2
     2.944033 Prior layer slice 3 Neuron 3
     3.762620 Prior layer slice 3 Neuron 4
    -1.695120 BIAS

Layer 6 of 6 (Full) Slice (this neuron) 1 of 6
         2.693996 Prior layer slice 1 Neuron 1
        -0.313751 Prior layer slice 2 Neuron 1
        -3.208661 Prior layer slice 3 Neuron 1
        -1.088728 Prior layer slice 4 Neuron 1
         0.714087 BIAS

...

Layer 6 of 6 (Full) Slice (this neuron) 6 of 6
        -1.245246 Prior layer slice 1 Neuron 1
        -4.326880 Prior layer slice 2 Neuron 1
         1.525335 Prior layer slice 3 Neuron 1
         1.519400 Prior layer slice 4 Neuron 1
        -1.512020 BIAS

The CUDA.LOG File

CONVNET also writes a file called CUDA.LOG. It is divided into four sections. The first section names the CUDA device present and lists its capabilities. The second section lists the architectural and training parameters given by the user. The third section shows the device memory allocations, along with some supplementary information about allocation of convolutional gradient scratch memory. This may be of interest if device memory is limited and the user needs to tweak parameters to make optimal use of memory.

The last section is the most useful. It shows the total and per-launch device time, broken down by layer and by activity in each layer (forward-pass activation, backpropagation of delta, and computation of gradient). It also lists several other CUDA-related activities.

What makes this table important is the per-launch times. Windows imposes a limitation on this time. Currently, the default limit is two seconds. It can be changed with a registry hack, but you won't hear about it from me. The key thing is that this per-launch time lets the user tweak parameters. If gradient computation is the dominant per-launch issue, then the "Max CONV work per launch" parameter can be reduced. If activations are also a problem, the "Max batch size" parameter can be reduced.

Index

© Timothy Masters 2018
T. Masters, *Deep Belief Nets in C++ and CUDA C: Volume 3*, https://doi.org/10.1007/978-1-4842-3721-2

S, T, U, V, W, X, Y, Z